DYNAMIC PROGRAMMING
MODELS AND APPLICATIONS

Eric V. Denardo
Professor of Operations Research
Yale University

DOVER PUBLICATIONS, INC.
Mineola, New York

Bibliographical Note

This Dover edition, first published in 2003, is an unabridged and slightly corrected republication of the work originally published by Prentice-Hall, Inc., Englewood Cliffs, N.J., in 1982. A new Preface to the Dover Edition has been added.

Library of Congress Cataloging-in-Publication Data

Denardo, Eric V., 1937–
    Dynamic programming : models and applications / Eric V. Denardo.
        p. cm.
    Originally published: Englewood Cliffs, N.J. : Prentice-Hall, c1982.
    Includes bibliographical references and index.
    ISBN 0-486-42810-9 (pbk.)
        1. Dynamic programming. I. Title.

T57.83 .D45 2003
519.7'03—dc21

2002041346

Manufactured in the United States by Courier Corporation
42810903
www.doverpublications.com

Final term exam
12. 13

To Richard E. Bellman,
the father of dynamic programming

Lei Zhao

IOE 512

CONTENTS

SUPPLEMENT 2

PREFACE TO THE DOVER EDITION

The original edition of this book, published in 1982, strove to provide a brief account of the basic ideas of dynamic programming. In the ensuing decades, dynamic programming has become more widely used in economics, finance, operations management, and elsewhere, and the power of digital computers has jumped a millionfold. For these reasons, a thoroughly up-to-date introduction to dynamic programming would differ a bit from this book.

And yet, the book seems to have had staying power. Why is that? Firstly, it strives to lay bare the central themes of dynamic programming as quickly and simply as possible. For instance, the book begins with the most rudimentary dynamic program, a shortest-path problem. As soon as that problem has been solved, it is used to introduce and illustrate *embedding*, the *functional equation*, and several versions of the *principle of optimality*. Second, this book stresses not merely the computation of optimal policies, but qualitative insight into their structure. This is true of structured shortest-path problems and of the "(s, S)-policies" for controlling inventory, to cite two examples. Third, when analyzing a particular type of problem, the book strives to use a method that applies more generally. For instance, to solve a discounted Markov decision problem, the book takes an approach that holds broadly, if contraction mappings are introduced. Fourthly, the problems at the end of each chapter bring the chapter's methods to bear on related problems and, by doing so, survey the related literature. In brief, the book strives to be a brief account of the key ideas of dynamic programming.

If I were to add two chapters to this book, one of them would build a general model of discounted dynamic programming, using contraction mappings. The other chapter would focus on models whose optimal policies have an interesting structure known as a "Gittins index" (after J. C. Gittins, who pioneered this area). In addition, I would also emphasize the formulation of dynamic programs for solution as a linear program (LP).

PREFACE

Nearly a quarter century has elapsed since the publication of Richard Bellman's path-breaking 1957 book, *Dynamic Programming*. His book and papers spurred rapid development of dynamic programming and of the portions of operations research in which it is used. Facets of this development have been explored in various texts, but this volume provides the first up-to-date account since Bellman's of the key ideas of dynamic programming and of their uses in operations research. All of the material presented here is basic, but little of it has appeared previously in book form.

The core of this volume is accessible to individuals who have had one prior course in operations research, including elementary probability. The author does not, however, assert that this volume is easy reading. As is common with "elementary" books, a great many ideas are introduced here. Certain optional starred sections do probe beyond the elementary level, and their prerequisites are clearly marked.

A significant fraction of the literature on operations research is presented in this volume's eight chapters and two supplements. Seven chapters discuss models that have finite planning horizons. The remaining chapter glimpses our companion volume, which treats models with indefinite and infinite planning horizons. The two "supplements" contain facts that should be known by every student of operations research and by every professional. One supplement concerns data structures, which are the ways in which information can be organized in computers. The other supplement contains a lucid account of the basic properties of convex functions.

Brief mention of this book's chapters reveals its scope. Chapter 1 is a layman's introduction to sequential decision processes, the things studied by dynamic programming. Chapter 2 focuses on the simplest sequential decision process, a shortest-route problem, and it surveys methods for computing shortest routes. Chapter 3 uses dynamic programming to study models of resource allocation, with emphasis on marginal analysis and including a modern account of Lagrange multipliers. Chapter 4 provides methods for approximating solutions to control problems in continuous time. Chapter 5 studies production control, with special emphases on economies of scale and on diseconomies of scale. Chapter 6 studies decision making in the face of an uncertain future. Chapter 7 analyzes inventory control models. It provides conditions under which (s, S)-policies are optimal, and it shows why these conditions obtain. Chapter 8 introduces sequential decision processes that lack fixed planning horizons. Each chapter contains basic material to which access had previously been difficult, if possible. The chapters are best read in numerical order, except that Chapters 7 and 8 are independent of each other.

This volume is liberally sprinkled with exercises, which are placed where working them would do the most good. Problem sets may be found at the end of each chapter. These problems review the ideas developed in the chapter, study related issues, and survey related literature. A number of these problems are new, and many of them can serve as material for classroom presentations and lectures.

Thanks are due to the many individuals who influenced this volume. Loring G. Mitten, the author's dissertation advisor, interested him in the subject and nurtured his early development. Helpful suggestions on various chapters have been made by Henk Tijms, Awi Federgruen, Ward Whitt, Salah E. Elmaghraby, and others. Bennett L. Fox has contributed greatly to the author's knowledge of this subject. He has read a great many drafts, and he has improved the content and presentation in a great many ways. Thanks are also due to the Center for Operations Research and Mathematical Economics (CORE), then in Louvain, Belgium, and to the Mathematical Institute at the University of Aarhus for providing the respites that helped bring this project to fruition. This volume also benefits from the reactions of the many students who have studied drafts. Daniel Luan helped compile the bibliography. Mrs. Marie Avitable typed the final draft and several of its precursors.

INTRODUCTION

TO SEQUENTIAL

DECISION PROCESSES

DECISION MAKING
DYNAMIC PROGRAMMING
BEGINNINGS
SCOPE
ADVICE TO THE READER

DECISION MAKING

Nearly every facet of life entails a sequence of decisions. Managing one's assets involves the evaluation of a never-ending stream of purchase, sales, and exchange opportunities. Managing a retail store involves a sequence of purchasing, pricing, and advertising decisions. Playing tennis entails a series of serves, returns, and volleys, all carefully interwoven. Driving to the store requires a sequence of accelerations, brakings, lane changes, and turns. Piloting a spaceship entails a sequence of maneuvers and course changes.

These sequential activities are ubiquitous, and they have some common features. Each has a purpose. An individual might drive with the intent of arriving quickly at his or her destination, without breaking the law. Many athletes compete for the satisfaction of winning. Business managers attempt to maximize profit.

In each of these activities there is interplay between constituent decisions. Making a purchase consumes capital and restricts future options. Igniting a spaceship's rocket consumes fuel that cannot be used later.

In many cases, decisions must be made without knowing their outcomes. A driver cannot predict how others will react to his or her moves; if drivers could predict reactions, there would be no accidents. A tennis player does not know in advance whether a particular shot will be in bounds, or how it will be returned. Investors and gamblers make choices without knowing outcomes. Indeed, investing, gambling, and sporting events would lose their fascination if their outcomes could be foretold.

The term *sequential decisions process* is used to describe an activity that entails a sequence of actions taken toward some goal, often in the face of uncertainty. The activities just discussed are all sequential decision processes, as are so many facets of modern life.

DYNAMIC PROGRAMMING

Sequential decision processes as diverse as playing tennis and managing inventory seem to have little in common. Yet their mathematical representations share several important features. In fact, the subject known as dynamic programming was born of the realization that certain features recur again and again when sequential decision processes are analyzed. *Dynamic programming* is the collection of mathematical tools used to analyze sequential decision processes. This identifies dynamic programming as a branch of applied mathematics rather than as something more specific. The subject's coherence results from the fact that it is pervaded by several themes. We shall see that these themes include the concept of states, the principle of optimality, and functional equations.

Dynamic programming can bear fruit in the form of insights or numbers. The insights usually come from theory, the numbers from computation. For instance, in an elementary inventory situation, one can determine that the optimal policy consists of drawing a line on a bin and refilling the bin each week to that line. This is the sort of insight that can be obtained by theory. It rules out more complex control policies. If one wishes to know where to draw the line, one must compute, usually on a digital computer.

We must note that some sequential decision processes lie beyond the reach of dynamic programming. Consider chess. This is a game of finitely many positions. A routine application of dynamic programming yields a system of one equation per position, and the solution to this equation system determines optimal play for both players. The trouble is that the number of equations (one per position) is astronomical, and there is virtually no hope that they will ever be solved on any computer.

BEGINNINGS

Dynamic programming is no exception to the rule that good ideas have deep roots. Ancestors to its functional equations can be traced through the history of mathematics. But the labors that give birth to the field of dynamic programming bear recent dates. They include Massé's study (1946) of water resources; Wald's study (1947) of sequential decision problems in statistics; a related study by Arrow, Blackwell, and Girshick (1949); studies of the control of inventories by Arrow, Harris, and Marshak (1951) and by Dvoretsky, Kiefer, and Wolfowitz (1952a, 1952b); and the study by Bellman (1952) of functional equations.

It was Bellman who seized upon the principle of optimality and, with remarkable ingenuity, used it to analyze hundreds of optimization problems in mathematics, engineering, economics, operations research, and other fields. Many of his contributions are compiled in his book on dynamic programming (1957a), others in his joint work with Dreyfus (Bellman and Dreyfus, 1962). Isaacs (1951), whose "tenet of transition" is another version of the principle of optimality, seems to merit a share of credit for recognizing the key role that this idea plays.

SCOPE

This volume is devoted to sequential decision processes that have definite "ends" at which decision making stops. These models are said to have *definite planning horizons*. They can be analyzed by induction, working backward from the end of the planning horizon to its beginning. Induction is the simplest mathematical tool, which might seem to connote that these models are low on intellectual fiber. Nothing of the sort—readers of this volume are offered plenty on which to chew.

A companion volume is devoted to sequential decision processes that have indefinite ends or, equivalently, *indefinite planning horizons*. That volume entails a higher level of mathematical sophistication, but no more ingenuity.

ADVICE TO THE READER

This volume is designed for both those who seek a passing acquaintance with dynamic programming and for those who wish to become expert. Elementary and advanced aspects of a subject are contained in the same chapter. The advanced material is confined to starred sections. It is recommended that most readers thumb through or skip the starred portions entirely, at least the first time through. Titles of starred sections are set in boldface type in each chapter's contents page.

Supplements 1 and 2 concern data structures and convexity, respectively. These supplements are used extensively in the starred sections, but not in the unstarred ones.

Exercises are scattered throughout this volume, and they are placed where working them would do the most good, which is often in the middle of a development. Pause to try the exercises. If one troubles you, it may indicate that you have missed something on the preceding pages. Look back for clues.

2

THE PROTOTYPE

SEQUENTIAL DECISION

PROCESS

Titles above in bold type concern advanced
or specialized subjects. These sections can
be omitted without loss of continuity.

OVERVIEW

Chapters 2 to 4 have two purposes. One is to study deterministic sequential decision processes, whose evolution is not affected by chance. The other is to acquaint you with the basic ideas that underlie and unify dynamic programming. Each of these ideas illuminates the others, but most beginners find all of them strange and alien.

The presentation reflects this state of affairs. Chapters 2 to 4 are more redundant than is usual in technical writing. This is intentional. It offers you multiple exposures to each basic idea. It allows you to think about each idea after having been exposed to the others.

The starred sections of each chapter contain more advanced topics that can be deferred or omitted.

DIRECTED NETWORKS

Sequential decision processes exist in many varieties, but all are elaborations of the simple prototype that is analyzed in this chapter. This prototype is an optimization problem in a directed network, a fact that obliges us to discuss networks first. However, the diversion pays dividends, as the notation, terminology, and pictographs of directed networks will prove useful time and again.

A directed network has an alluringly simple pictorial representation, but a slightly abstruse mathematical definition. Mathematically, a *directed network* consists of a nonempty set S of *nodes* and a subset T of the Cartesian product

$S \times S$. The elements of T are called *directed arcs*. This means that a directed arc is an ordered pair (i, j), where i and j are nodes. To see the picture, think of nodes i and j as circles with the symbols i and j inside, and think of directed arc (i, j) as a line connecting these circles, with an arrow pointing from i to j. Figure 2-1 depicts a network that has five nodes and six directed arcs. The nodes are numbered 1 through 5, and (3, 2) is a directed arc, but (2, 3) is not.

FIGURE 2-1.

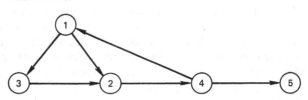

This paragraph contains a dozen definitions that are easily remembered, as each is suggested by normal English usage. The motivating idea is that directed arc (i, j) suggests the possibility of direct *movement* from node i to node j. The term "path" is intended to connote a sequence of such movements. Mathematically, a *path* is a sequence $(i_1, i_2, \ldots, i_n)$ of at least two nodes that has the property that (i_k, i_{k+1}) is a directed arc for $k = 1, 2, \ldots, n - 1$. The illustrated network contains paths (2, 4) and (1, 3, 2) but not (1, 2, 3). Directed arc (i, j) is said to *emanate* from node i and *terminate* at node j. Path $(i_1, \ldots, i_n)$ is called a path *from* node i_1 *to* node i_n. This path is said to *contain* path $(i_p, \ldots, i_q)$ whenever $1 \leq p < q \leq n$. As an example, path (1, 2, 4, 1) contains itself and several other paths, including (1, 2, 4) and (2, 4). A path with $i_1 = i_n$ is called a *cycle*. Path (1, 2, 4, 1) is a cycle in the illustrated network. A network is called *cyclic* if it contains at least one cycle. A network that contains no cycle is called *acyclic*. The illustrated network is cyclic, but would be acyclic if directed arc (2, 4) were removed or reversed. Those nodes from which no arcs emanate are called *ending* nodes; those at which no arcs terminate are called *beginning* nodes. Node 5 is an ending node. The network in Figure 2-1 has no beginning nodes; it would if directed arc (2, 4) were deleted. *Finite* networks have finitely many nodes. We often abbreviate "directed network" to "network" and "directed arc" to "arc." This should not cause confusion, because we never discuss networks whose arcs lack directions.

Optimization problems emerge when a *length* t_{ij} is associated with arc (i, j). Normally, the *path length* is defined as the sum of the lengths of the path's arcs. A typical optimization problem is to find the longest (or shortest) path from one node to another—or from one node to every other node—or between every pair of nodes. Another type of optimization problem is to find the cycle having largest (or smallest) ratio of length to number of arcs. These optimization problems can be treated effectively by the methods of dynamic programming.

The networks studied in the unstarred sections of this chapter are particularly simple. They are finite and acyclic. A key property of such networks is introduced in terms of Figure 2-2. This network is finite and acyclic, but the labels have purposely been omitted from its nodes. These nodes can be labeled with the integers 1 through 4 in such a way that each arc (i, j) has i less than j. Do so. (There is exactly one such labeling.) Labeling Figure 2-2 in this way should convince you that any acyclic network consisting of N nodes can have its nodes labeled with the integers 1 through N so that each arc (i, j) has i less than j. For the cognoscenti, we now sketch a labeling procedure that accomplishes this. Note that at least one node must have no arcs terminating at it (else there would be a cycle). Label this node with the integer 1. Then delete it and all arcs emanating from it. Examine the remaining network. Note that it must have at least one node at which no arcs terminate (for the same reason). Label that node with the integer 2. Et cetera.

FIGURE 2-2.

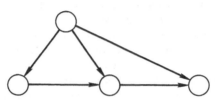

SHORTEST AND LONGEST PATHS

The prototype of all dynamic programming problems is to find the shortest (or the longest) path through a finite acyclic network. In acyclic networks, shortest-path and longest-path problems are interchangeable; either can be converted to the other by multiplying the lengths of all the arcs by -1. A shortest-path problem is now worked out in complete detail. Then a longest-path problem is analyzed, with some of the details left to you.

Shortest-Route Problem

John bicycles to work regularly. Being methodical, he has decided to calculate the fastest of several routes from his home to the office. With some effort, he has collected the data summarized in Figure 2-3. Node 1 represents his home, node 9 represents his office, and the remaining nodes represent road junctions. Directed arcs represent roads between junctions. The number adjacent to each arc is the travel time in minutes on the road it represents; it takes 15 minutes to travel from junction 4 to junction 7. The arrow is a somewhat artificial device; it represents John's prior conviction that when bicycling to work he might wish to traverse a road in the direction of the arrow, but would never wish to do so in the opposing direction. Effectively, all the "streets" are

FIGURE 2-3.

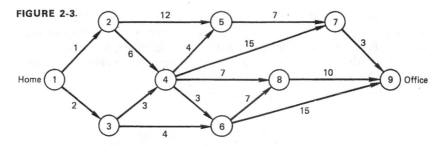

one-way; a version with two-way streets is described in the chapter's starred section. Note that the nodes are labeled such that each arc (i, j) has $i < j$. Note also that the labeling is not unique (e.g., labels 2 and 3 can be interchanged).

With a cursory examination of Figure 2-3, you can satisfy yourself that the fastest path from home to office is $(1, 3, 4, 5, 7, 9)$ and that its travel time is 19 minutes. Having solved John's problem by inspection, we now do so by dynamic programming. The intent of this is to provide insight into dynamic programming. Let

$$f_i = \text{the minimum travel time from node } i \text{ to node } 9$$

By definition, f_9 is zero, and you have just observed that $f_1 = 19$. Let t_{ij} denote the travel time along arc (i, j), so $t_{47} = 15$. Interpret

$$t_{ij} + f_j$$

as the travel time of the path from node i to node 9 that first traverses arc (i, j) and then travels as quickly as possible from node j to node 9. As this is a path from node i to node 9, its travel time must be at least as large as f_i. In other words,

$$f_i \le t_{ij} + f_j, \qquad i \ne 9$$

But the fastest path from node i to node 9 traverses *some* arc (i, j) first and then gets from node j to node 9 as quickly as possible. So some j satisfies the displayed inequality as an equality, and

(2-1) $$f_i = \min_j \{t_{ij} + f_j\}, \qquad i \ne 9$$

The set of those j over which the right-hand side of equation (2-1) is to be minimized has not been represented explicitly; minimization occurs over those j for which (i, j) is an arc. When $i = 4$, one minimizes the right-hand side over the values 5, 6, 7, and 8 of j. The case $i = 9$ is not covered by equation (2-1), but we have already observed that

(2-2) $$f_9 = 0$$

Notice that equation (2-1) specifies f_i once f_j is known for every j such that (i, j) is an arc. The nodes in Figure 2-3 have been numbered so that every arc (i, j) has j greater than i. Hence, f_i can be determined from (2-1) once f_j is known for every j greater than i. Consequently, (2-1) allows computation of

f_i in *decreasing* i, starting with $i = 8$ and ending with $i = 1$. If this is your first exposure to dynamic programming, you are advised to find pencil and paper, perform the somewhat tedious calculation just described, and compare your results with what follows. (Alternatively, you might do the calculation directly on Figure 2-3.)

$$f_8 = 10 + f_9 = 10$$

$$f_7 = 3 + f_9 = 3$$

$$f_6 = \min \begin{cases} 7 + f_8 \\ 15 + f_9 \end{cases} = \min \begin{cases} 7 + 10 \\ 15 + 0 \end{cases} = 15$$

$$f_5 = 7 + f_7 = 10$$

$$f_4 = \min \begin{cases} 4 + f_5 \\ 15 + f_7 \\ 7 + f_8 \\ 3 + f_6 \end{cases} = \min \begin{cases} 4 + 10 \\ 15 + 3 \\ 7 + 10 \\ 3 + 15 \end{cases} = 14$$

$$f_3 = \min \begin{cases} 3 + f_4 \\ 4 + f_6 \end{cases} = \min \begin{cases} 3 + 14 \\ 4 + 15 \end{cases} = 17$$

$$f_2 = \min \begin{cases} 12 + f_5 \\ 6 + f_4 \end{cases} = \min \begin{cases} 12 + 10 \\ 6 + 14 \end{cases} = 20$$

$$f_1 = \min \begin{cases} 1 + f_2 \\ 2 + f_3 \end{cases} = \min \begin{cases} 1 + 20 \\ 2 + 17 \end{cases} = 19$$

Figure 2-4 summarizes the information obtained from this calculation. Travel time f_i is recorded above node i, each arc (i, j) attaining the minimum in (2-1) is preserved, and all other arcs are deleted. Note that Figure 2-4 prescribes the fastest path from *each* node to node 9, not just from node 1 to node 9. It turned out to be easiest to solve John's problem by embedding it in the more general problem of finding the shortest path from each node to node 9.

Equation (2-1) is the prototype of the equations of dynamic programming. The argument used to justify (2-1) holds whether or not the network's nodes

FIGURE 2-4.

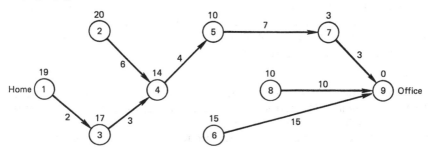

are labeled so that each arc $(i\ j)$ has $i < j$. It holds for any arc lengths, including negative lengths. It even holds for cyclic networks, provided that no cycle has negative length.

Although (2-1) holds more generally, the preceding computational method works only for those acyclic networks whose nodes are labeled so that each arc (i,j) has $i < j$. This labeling often arises naturally, as when decision making evolves over time. More general methods of computation are described in this chapter's starred section.

Sequential decision processes usually represent decision problems involving trade-offs between immediate costs and future costs. The *myopic solution* to the shortest-route problem selects for each node i the shortest directed arc (i, j) emanating from it. The myopic solution disregards the node at which an arc terminates, for which reason it may be far from optimal. In fact, the network in Figure 2-3 is designed so that the myopic solution to the shortest-route problem yields the longest path from each node to node 9.

When an equation system has exactly one solution, it is said to *characterize* that solution. The numbers $\{f_1, \ldots, f_9\}$ satisfy (2-1) and (2-2). Moreover, we have computed these numbers from (2-1) and (2-2), which means that no other numbers satisfy these equations. In other words, (2-1) and (2-2) *characterize* the set $\{f_1, \ldots, f_9\}$ of fastest travel times to node 9. [In this, we mean that the functional values are unique, not that the maximizing indices in (2-1) are unique; ties occur when multiple paths have the same length.]

Shortest-Path Trees

The network displayed in Figure 2-4 has one fewer arc than node, and it prescribes exactly one path from each node to node 9. This network is a *tree* of paths to node 9. The tree's path from node i to node 9 has length f_i, which is the shortest of the lengths of all the paths from node i to node 9. Hence, the network depicted in Figure 2-4 is a tree of shortest paths, or a *shortest-path tree*. Rather than finding the shortest path from node 1 to node 9, we have found a tree of shortest paths to node 9 from all other nodes.

More generally, a subset of the arcs in a network is called a *tree* of paths *to node j* if this arc set contains exactly one path from each node i, with $i \neq j$, to node j. Similarly, a subset of arcs is called a *tree* of paths *from node i* if this arc set contains exactly one path from node i to each node j, except $j = i$. No tree can contain a cycle, because then it would have multiple paths. One can show that every tree contains one fewer arc than the number of nodes in the network. When the methods of dynamic programming are applied to network optimization problems, one often gets a tree of shortest paths.

Longest-Route Problem

One might suspect that only extreme pessimists could be interested in the longest path through a directed acyclic network, but this will turn out to be false. Consider the problem of finding the longest path from node 1 of Figure

2-3 to node 9. Partly for variety, we now investigate paths from node 1 to node j, rather than from node j to node 9. With f_1 set equal to 0, let

f_j = the length of the longest path from node 1 to node j

The inequality

$$f_j \geq f_i + t_{ij}$$

reflects the fact that $f_i + t_{ij}$ is the length of a path from node 1 to node j. The longest path from node 1 to node j has final arc (i, j) for some i. So, for some i, the preceding inequality holds as an equality; that is, with $f_1 = 0$,

(2-3) $$f_j = \max_i \{f_i + t_{ij}\}, \quad j > 1$$

where maximization occurs over all i such that (i, j) is an arc. All arcs point from lower-numbered nodes to higher-numbered nodes. So, the right-hand side of (2-3) can be evaluated for f_j successively with $j = 2$, then 3, and so on, ending with $j = 9$. Doing so produces the information in Table 2-1, which records for each j the path length f_j and the index i such that (i, j) is the final arc in this path.

TABLE 2-1. Solution of a longest-route problem

Node j	1	2	3	4	5	6	7	8	9
f_j	0	1	2	7	13	10	22	17	27
maximizing i	—	1	1	2	2	4	4	6	8

Exercise 1. Check the computation in Table 2-1. Trace back the "maximizing i" to reconstruct the path (1, 2, 4, 6, 8, 9) whose length is 27. (Tracing back an optimal path is appropriately called *backtracking*.)

The "maximizing i" in Table 2-1 give the following set of eight arcs:

(1, 2), (1, 3), (2, 4), (2, 5), (4, 6), (4, 7), (6, 8), (8, 9)

These eight arcs form a *tree of longest paths* from node 1 to all other nodes in this nine-node network.

The Critical Path

For an illustration of the longest-route problem, reconsider Figure 2-3. Imagine that John is building a house. This entails many activities, such as laying a foundation, framing, plumbing, wiring, and so on. Some of these activities can be done concurrently, and others cannot. The question is: How quickly can the house be built?

In our model of this problem, each directed arc depicts a particular activity, such as hiring a contractor or laying a foundation. The number associated with directed arc (i, j) is the number of weeks it takes to complete activity (i, j). The nodes represent the fact that certain activities cannot be started until others are completed. Specifically, node i depicts the fact that each activity (arc) (i, j)

emanating from node i can only be started after all activities (k, i) terminating at node i have been completed. One can start the four activities whose arcs emanate from node 4 as soon as one completes the two activities whose arcs terminate at node 4. Node 1 represents the start of the project, and node 9 represents its completion.

How many weeks does it take to complete the project? Let it commence at week 0. A path from node 1 to node i represents a sequence of activities that must be completed before activity (i, j) can commence. So, activity (i, j) can commence no earlier than week f_i, which is the length of the longest path from node 1 to node i. It can be shown, by induction, that activity (i, j) can commence at week f_i and not earlier. This means that the project completion time is the length of the longest path from node 1 to node 9, which is 27 weeks. The longest path $(1, 2, 4, 6, 8, 9)$ from node 1 to node 9 is called the *critical path*. Its arcs depict the activities warranting closest coordination.

Exercise 2. Interpret (2-1) with "min" replaced by "max". Use it to compute the latest times at which activities might commence while not increasing the project completion time.

This discussion leaves open the question of how to construct a network whose longest path is the critical one. That turns out to require, in the most general cases, either zero-length arcs or multiple arcs per activity. Problem 4 includes a recipe having zero-length arcs.

SOME RECURRING FEATURES

The longest-route and shortest-route problems illustrate several features that are shared by a great many sequential decision processes. We shall now discuss four of these features: *embedding*, the *functional equation*, *recursive fixing*, and the *principle of optimality*. Other features are discussed in subsequent chapters, after more examples are introduced.

Interest in the shortest- and longest-route problems lay initially in the computation of f_1 and of a path whose length is f_1. Yet the problem of computing f_1 was not attacked directly. Instead, this problem was first embedded in a class of optimization problems—namely, the computation of f_i for each node i. Virtually every model of a sequential decision process entails an embedding of this type. So we display embedding as one of the common features.

> The problem of interest is *embedded* in a set of optimization problems, one problem per node (state).

In Chapter 3 "node" will be displaced by the more common term "state," which is placed in parentheses in the statement above.

In many sequential decision processes, the decision maker is really concerned with only one of the optimization problems in this set. So embedding seems a step backward, as it replaces one problem with many. However, the next two features will combine to prescribe a particularly efficient method for solving every problem in the set.

> The set of solutions to the optimization problems is characterized by a *functional equation*, such as (2-3) or (2-1) and (2-2).

The displayed statement means, for instance, that the set of quickest travel times to node 9 satisfies (2-1) and (2-2), but no other numbers do. Equations such as (2-1) and (2-2) are almost invariably called functional equations, even though this term reveals little about the structure of the equations.[1] For present purposes, think of a functional equation as a system of equations, one per node, that interrelate the solutions to the several optimization problems; moreover, most equations in the set have an optimization operator such as *max, min, minimax,* or *maximin* to the right of the equality sign, with variables such as f_i and f_j on both sides.

Functional equations (2-1) and (2-2) characterize the set of solutions to the optimization problems, and we solved them by evaluating (2-1) in decreasing i. Similarly, we solved functional equation (2-3) by evaluating it in increasing j. In both cases, the functional equation was solved by fixing the f-values in a preset sequence, as described below.

> The solution to the functional equation can be computed by a process called *recursive fixing*, which consists of evaluating its equations for the various nodes (states) in a predetermined sequence.

The term "recursive fixing" is intended to mean that the f-values are fixed (made permanent) one after another, in a certain preset sequence. Recursive fixing works for finite acyclic networks. One starts at one end and works toward the other, fixing f-values *en route*.

In some applications, the nodes in the network reflect calendar times, and the arcs reflect increases in calendar time. Such networks are necessarily acyclic; a cycle would entail one or more time reversals. When arcs represent the passage of time, functional equations such as (2-1) and (2-2) solve the family of *initial-value* problems, and functional equations such as (2-3) solve the family of *final-value* problems.[2]

Note that recursive fixing entails *exactly one* evaluation per arc. For instance, in the shortest-route problem, the evaluation for arc (i, j) consists of

---

[1] The term "functional equation" is well established in the literature, although the names *characteristic equation, optimality equation,* and *extremal equation* all seem (to the author) more descriptive of (2-1) and (2-2). A functional (i.e., a function that assigns a real number to each state) is present, so that the term passes muster from a technical viewpoint. But identifying the functional at this (or any) stage of the development serves little purpose.

[2] Recursive fixing has other names. Some writers use the term *backward optimization* for solution of initial-value problems by recursive fixing and the term *forward optimization* for solution of final-value problems by this method. A few writers call recursive fixing *dynamic programming*.

adding t_{ij} to f_j and comparing the sum with the best estimate of f_i found so far. One can hardly hope for a more efficient computational procedure than this, *provided* that the arcs of the network are unrelated in length.

Of the general features now being abstracted from the shortest-route problem, the most elusive is the so-called "principle of optimality," which has several useful versions. A version particularly well suited to networks employs this definition: an *optimal path* from node i to node j is the shortest (or longest when maximizing) of the paths from node i to node j. Recall that path $(i_1, \ldots, i_n)$ contains path $(i_p, \ldots, i_q)$ whenever $1 \leq p < q \leq n$.

> *Principle of optimality* (first version): Consider an optimal path from some node to some other node. Any path $(i_p, \ldots, i_q)$ contained in this path is an optimal path from node i_p to node i_q.

To illustrate, we recall from Figure 2-4 that the optimal path from node 1 to node 9 is $(1, 3, 4, 5, 7, 9)$. The principle of optimality asserts that since path $(3, 4, 5, 7)$ is contained in an optimal path, it is an optimal path from node 3 to node 7. If you wish you might check this assertion by turning back to Figure 2-3. But you need not flip back, for if another path p from node 3 to node 7 were shorter, then the composite path $(1, p, 9)$ would be shorter than the shortest path, $(1, 3, 4, 5, 7, 9)$, and this is impossible.

The principle of optimality and the functional equation are intimately interrelated. In many situations, each implies the other. In particular, the argument just used to justify the principle of optimality contained the key to deriving functional equations (2-1)-(2-2), and (2-3). However, this correspondence is imperfect, as Problem 7 illustrates by redefining the path length as the maximum of the arc lengths rather than their sum. Problem 6 observes that this version has an analogue in the calculus of variations.

Regrettably, the first version of the principle of optimality is insufficiently general. A widely applicable version entails a definition. In the context of the shortest-route problem in Figure 2-3, a *policy* specifies for each node i, with $i \neq 9$, a directed arc (i, j) emanating from node i. (Node 9 is special in that no arcs emanate from it.) This example has 64 different policies. Each policy specifies a tree of paths to node 9. A policy is called *optimal for node i* if the arcs it designates include an optimal path from node i to node 9. Clearly, a policy can be optimal for one node but not for another. To be optimal for node i, need a policy be nonoptimal for other nodes? No. The shortest-path tree in Figure 2-4 represents a policy that is optimal for every node! A policy that is optimal for *every* node is called an *optimal policy*. The second version of the principle of optimality asserts the existence of such a policy.

> *Principle of optimality* (second version): There exists a policy that is optimal for every node (state).

An advantage of the second version is that it nearly always holds. The first two versions are equivalent to each other for most network models. However, Problem 7 satisfies the second version but not the first.

The shortest-path tree in Figure 2-4 consists of the decisions in an optimal policy. Also, the longest-path tree, whose arcs are given below Table 2-1, represents an optimal policy. In network problems, optimal policies often turn out to be trees of shortest or longest paths.

Finally, we discuss the traditional version of the principle of optimality. This version concerns deviations from the optimal policy. In these deviations, the first arc (decision) can be changed, but all other arcs (decisions) are held constant.

> *Principle of optimality* (third version): An optimal policy has the property that whatever the initial node (state) and initial arc (decision) are, the remaining arcs (decisions) must constitute an optimal policy with regard to the node (state) resulting from the first transition.

The third version is intimately related to the functional equation; that is, the sum $t_{ij} + f_j$ that appears in functional equation (2-1) is the length of the path that chooses arc (i, j) initially and thereafter uses the arcs in an optimal policy.

It would be hard to overstate the importance of the principle of optimality. One of Richard Bellman's deepest insights has been, perhaps, that the principle of optimality is shared by a host of optimization problems whose mathematical formulations are so very disparate. The term *principle* of optimality is, however, somewhat misleading; it suggests that this is a fundamental truth, not a consequence of more primitive things. Several traditional developments of dynamic programming accept it as a principle and derive from it algorithms that compute optimal policies. The approach taken here is to recognize the principle of optimality as a common theme that guides us as we derive functional equations from more primitive properties of optimization problems.

REACHING

Recursive fixing is a way of computing solutions to longest-route and shortest-route problems in acyclic networks. These problems can be solved by a different (and sometimes faster) technique called *reaching*. Reaching is introduced by contrast with recursive fixing.

Consider the finite acyclic network whose node set consists of the integers 1 through N and whose arc set consists of all pairs (i, j), where i and j are integers having $1 \leq i < j \leq N$. As before, arcs point to higher-numbered nodes. Arc (i, j) has length t_{ij}, where $-\infty \leq t_{ij} < \infty$. One wishes to compute, for $j = 2, \ldots, N$, the length f_j of the longest path from node 1 to node j. Interpret an arc whose length equals $-\infty$ as an "uninteresting" notational convenience; f_j exceeds $-\infty$ if there exists at least one path of "interesting" arcs from node 1 to node j.

The longest path from node 1 to node j has some final arc (i, j); this arc (in fact, every arc) has $i < j$. Hence, with $f_1 = 0$,

$$(2\text{-}4) \qquad f_j = \max_{i | i < j} \{f_i + t_{ij}\}, \qquad j = 2, \ldots, N$$

Recursive fixing computes the right-hand side of (2-4) in ascending j, as specified below. (The symbols "DO" and "←" are made precise in the ensuing note.)

Recursive-fixing method: ∨

    1. Set $v_1 = 0$ and $v_j = -\infty$ for $j = 2, \ldots, N$.
    2. DO for $j = 2, \ldots, N$.   ∨
        3. DO for $i = 1, \ldots, j - 1$.

$$(2\text{-}5) \qquad v_j \longleftarrow \max\{v_j, v_i + t_{ij}\}$$

Note: In an algorithm, a variable can take a sequence of values. The expression $x \longleftarrow y$ means that x is to assume the value now taken by y. So (2-5) replaces v_j by $v_i + t_{ij}$ whenever the latter is larger. A "DO" statement means that what follows is to be executed for the specified sequence of values. So step 2 means that step 3 is to be executed $N - 1$ times, first with $j = 2$, then with $j = 3$, and so on. The "nest" of DO loops in steps 2 and 3 executes (2-5) in ascending j and, for each j, in ascending i. This fixes v_j equal to f_j in ascending j.

A different method consists of "reaching out" from a node i whose label v_i equals f_i to update labels v_j for all $j > i$.

Reaching method:

    1. Set $v_1 = 0$ and $v_j = -\infty$ for $j = 2, \ldots, N$.
    2. DO for $i = 1, \ldots, N - 1$.   ✓
        3. DO for $j = i + 1, \ldots, N$.

$$(2\text{-}6) \qquad v_j \longleftarrow \max\{v_j, v_i + t_{ij}\}$$

Expressions (2-5) and (2-6) are identical, but the DO loops are interchanged. Reaching executes (2-6) in ascending i and, for each i, in ascending j. One can show, by induction, that the effect of i executions of the nest of DO loops is to set each label v_k equal to the length of the longest path from node 1 to node k whose final arc (n, k) has $n \le i$. As a consequence, reaching stops with $v_k = f_k$ for all k.

One of the best ways to learn an algorithm is to work an example. Repeated below is our nine-node network, which now displays the status of reaching after three complete executions of the loop that consists of steps 2 and 3. The current value of label v_j is written above node j. The exercise that follows Figure 2-5 asks you to complete the calculation.

Exercise 3. Complete the application of reaching to the longest-route problem in Figure 2-5.

FIGURE 2-5.

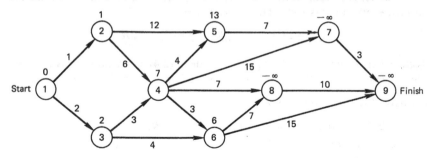

One wishes to determine a longest path, not just its length. Neither recursive fixing nor reaching has been specified in a way that facilitates this. To do so, record at node j the integer k for which arc (k, j) caused the most recent increase in v_j. This produces a tree of longest paths from node 1 to all others. It avoids any recalculation of f-values while backtracking a longest path.

The (primitive) versions of recursive fixing and reaching just given require work (number of computations) proportional to N^2. For sparse or structured networks, this chapter's final section gives faster methods.

WHEN REACHING RUNS FASTER

Reaching and recursive fixing entail the same calculations. That is, (2-5) is identical to (2-6), and each is executed exactly once per arc in the network. Moreover, when (2-5) or (2-6) is executed for arc (i, j), label v_i is known to equal f_i, but f_j is not yet known.

The key difference between reaching and recursive fixing is that i varies on the outer loop of reaching, but on the inner loop of recursive fixing. We now describe a simple (and hypothetical) situation in which reaching may run faster. Suppose it is known, prior to the start of calculation, that the f-values will be properly computed even if we omit executing (2-5) or (2-6) for all arcs (i, j) having $f_i > 10$. To exclude these arcs when doing reaching, add to step 2 a test as to whether v_i exceeds 10. If it does, omit step 3. This test is on the *outer* loop; it gets executed roughly N times. To exclude these arcs when doing recursive fixing, add to step 3 a test as to whether v_i exceeds 10. If it does, omit (2-5). This test is necessarily on the *inner* loop. It gets executed roughly $N^2/2$ times. [The exact numbers of tests are $(N-1)$ and $(N-1)N/2$, respectively, and the latter is larger for all $N > 2$.]

The following paragraphs contain four optimization problems whose structure accords reaching an advantage. In each case, this advantage stems from the fact that reaching puts the node whose f-value is known on the outer loop.

First, suppose it turns out that no finite-length path exists from node 1 to certain other nodes. One has $v_j = f_j = -\infty$ for these nodes. It is not necessary to execute (2-5) or (2-6) for any arc (i, j) having $v_i = -\infty$. If reaching is used, fewer tests are needed, because i varies on its outer loop.

Second, consider a shortest-path problem in which one is solely interested in f_k for a particular k, and suppose that all arc lengths are nonnegative. One need not execute (2-5) or (2-6) for any arc (i, j) having $v_i \geq v_k$. Checking for this is quicker with reaching, because i varies on its outer loop.

Third, consider a shortest-route problem in a directed acyclic network that has node 1 as its only beginning node, but does *not* have its nodes numbered so that each arc (i, j) has $i < j$. One could relabel the nodes as indicated above Figure 2-2 and then use recursive fixing. The number of computer operations (arithmetic operations, comparisons, and memory accesses) needed to relabel the nodes is, however, roughly the same as the number of computer operations needed to compute the shortest-path tree. A saving can be obtained by combining relabeling with reaching as follows. For each node i, initialize label $b(i)$ by equating it to the number of arcs that terminate at node i. Maintain a list L consisting of those nodes i having $b(i) = 0$. Node 1 is the only beginning node, so $L = \{1\}$ initially, Remove any node i from L. Then, for each arc (i, j) emanating from node i, execute (2-6), subtract 1 from $b(j)$, and then put j into L if $b(j) = 0$. Repeat this procedure until L becomes empty.

Fourth, consider a model in which node i describes the state of the system at time t_i. Suppose that each arc (r, s) reflects a *control* c_{rs} which is in effect from time t_r to time t_s, with $t_r < t_s$. Suppose it is known a priori that a monotone control is optimal [e.g., an optimal path $(1, \ldots, h, i, j, \ldots)$ has $c_{hi} \leq c_{ij}$]. This would occur, for instance, in an inventory control model where the optimal stock level is a monotone function of time. The optimal path from node 1 to node i has some final arc (h, i). When reaching is used, one need only execute (2-6) for those arcs (i, j) having $c_{hi} \leq c_{ij}$. The effect of this is to *prune* the network during computation on the basis of structural information. This is accomplished more quickly with reaching because i varies on its outer loop.

The main disadvantage of reaching is that it can entail more memory accesses; see Problem 8 for details.

SUMMARY

The prototypical model of a sequential decision process is a shortest (or longest)-route problem in a directed acyclic network. We have studied this problem with an eye to revealing the features that hold for more elaborate models. These features include embedding, the principle of optimality, functional equations, and recursive fixing. We also introduced a second computational method, called reaching, that proves to be faster than recursive fixing for certain structured models. That reflects a different theme of dynamic programming, which is to exploit structure to get qualitative results or to streamline computation.

In the final, starred section of this chapter, reaching will be adapted to solve shortest-route problems in cyclic networks.

SHORTEST PATHS IN CYCLIC NETWORKS*

The networks considered in this section need not be acyclic. Except for the discussion of label correction, all arcs have nonnegative lengths. Consider a directed network whose node set consists of the integers 1 through N and whose arc set consists of all ordered pairs (i, j) of integers i and j between 1 and N, inclusive. Each directed arc (i, j) has length t_{ij}, where $0 \leq t_{ij} \leq +\infty$.

The problem addressed here is to find the tree of shortest paths from node 1 to all others. Define f_j for $j = 2, \ldots, N$ by

$$f_j = \text{the length of the shortest path from node 1 to node } j$$

Self-loop arcs of the form (i, i) can be ignored because arc lengths are nonnegative. With $f_1 = 0$, one gets the familiar functional equation

$$(2\text{-}7) \qquad f_j = \min_{i | i \neq j} \{f_i + t_{ij}\}, \qquad j \neq 1$$

Exercise 4. Justify (2-7). Under what circumstance is f_j infinite?

The network is cyclic. Recursive fixing does not apply; that is, it is not possible to compute f_j in a preset sequence, such as ascending or descending j. Instead, we use reaching to compute f_j in nondecreasing f-value sequence, as follows.

Reaching (for cyclic networks having nonnegative arc lengths):

   1. Set $v_1 = 0$; set $v_j = t_{1j}$ for $j = 2, \ldots, N$, and set $T = \{2, \ldots, N\}$.

   2. Find a node i in T for which

$$(2\text{-}8) \qquad v_i = \min \{v_j | j \text{ in } T\}$$

   3. Delete i from T. Stop if T is (now) empty. Otherwise, go to step 4.

   4. For each j in T, set

$$(2\text{-}9) \qquad v_j \longleftarrow \min \{v_j, v_i + t_{ij}\}$$

   5. Go to step 2.

Note: The nodes in T have *temporary* labels. Step 2 finds a node i whose label is smallest among the nodes having temporary labels. Step 3 declares node i's label *permanent*, and step 4 executes (2-9) for those nodes j remaining in T. To execute (2-9) efficiently on a computer, keep T in contiguous memory cells, as described in Problem 19.

---

* This section concerns a special topic that can be omitted without loss of continuity. This section uses lists, heaps, and buckets, which are described in Supplement 1.

THEOREM 1. Suppose that $t_{ij} \geq 0$ for all (i, j). Reaching terminates with $v_j = f_j$ for each j.

Proof. Adopt the following inductive hypothesis. At any execution of step 2, v_j equals f_j for each j not in T; also, for each j in T, v_j equals the length of the shortest path from node 1 to node j none of whose intermediate nodes are in T. (Step 1 assures that the inductive hypothesis holds at the first execution of step 2.) Now consider the node i selected in step 2. Every path from node 1 to any node in T must have a first node j in T and, since arc lengths are nonnegative, must be at least as long as v_j. Since $v_j \geq v_i$, we are assured that $v_i = f_i$. In step 3, node i is deleted from T. In step 4, arcs (i, j) with j in T are evaluated, and the inductive hypothesis is restored for the revised (smaller) set T. ∎

Step 4 of reaching can be augmented to record, for each j, the arc giving the most recent reduction of v_j. This produces a tree of shortest paths from node 1 to all others.

Exercise 5. Ignore the arrows in Figure 2-3, allowing arc (i, j) to be traversed in both directions. Adapt reaching to compute the lengths of the shortest paths from node 4 to all others, and use it to compute these numbers.

A measure of the merit of an algorithm is the amount of *work* needed to execute it on a digital computer, where each computer operation (memory access, arithmetic operation, comparison, etc.) is counted as one unit of work. One can show that the preceding version of reaching takes work proportional to N^2. Other algorithms are proposed below, each with its work bound.

A network that has N nodes and an average of n arcs emanating from each node is called *sparse* if n is a small fraction of N. Many applications give rise to networks that are sparse. Applications having larger numbers of nodes tend to have sparser networks (i.e., n/N tends to decrease as N increases).

Consider what happens when reaching is implemented on a large sparse network. Suppose, for this discussion, that $N = 1000$ and that $n = 10$. So n/N equals 0.001. A table of all t_{ij} has N^2 (1 million) entries, of which only 0.1% (10,000) are finite. A more efficient data structure records only the finite entries, as follows.

Lists: For each node i, make a list of those j such that t_{ij} is finite. Execute (2-9) for each j on node i's list, rather than for each (i, j) having j in T.

Originally, (2-9) was executed for every j in T. There are, on the average, $N/2$ (500) such j. As just adapted, (2-9) is executed an average of only n (10) times per iteration.

In step 2, the smallest of $\{v_j | j \in T\}$ must be found. If one does this by comparison of v-values, it takes one fewer comparison than the number of elements in T, roughly $N/2$ (500) comparisons, on average. On average, at most

n (10) alterations to $\{v_j | j \in T\}$ occur per iteration. If can be more efficient to maintain these data in a heap, as follows.

Heaps: Maintain $\{v_j | j \in T\}$ as a heap with the smallest element on top. Exclude from the heap all elements j for which v_j is infinite.

The amount of work needed to execute reaching with lists and heaps is, in the worst case, proportional to $Nn \log_2 N$. Comparing this with the work bound of N^2 for the basic method suggests that the basic method is better for dense networks, but worse for sparse ones. Empirical evidence indicates, however, that reaching with heaps and lists is faster than the basic method for sparse networks and, surprisingly, competitive with it for dense ones.

The implementations with and without heaps must compete with the following simple method.

Label correction method:

    1. Set $v_1 = 0$, set $v_j = \infty$ for $j = 2, \ldots, N$, and set $L = \{1\}$.

    2. Stop if L is empty. Otherwise, delete a node i from L.

    3. For each arc (i, j) emanating from node i, execute (2-9) and, if v_j gets decreased, add j to L. Go to step 2.

One can show (see Problem 20) that, if the network has no cycle whose length is negative, label correction terminates finitely, with $v_j = f_j$ for each node j. Label correction works when some arcs have negative lengths. A second advantage of label correction is that it can be adapted (see Problem 21) to determine whether the network has a cycle whose length is negative and, if it does, to find such a cycle.

The computational efficiency of label correction depends on how one selects the item i to delete from L. One could keep L as a FIFO (first-in, first-out) list or as a LIFO (last-in, first-out) list. Label correction reduces to reaching if L is kept as an *ordered* list (i.e., if one deletes from L an item i whose v-value is smallest). The work needed to execute label correction with L kept as a FIFO list is proportional, at worst, to nN^2. With a LIFO list, it can be as bad as 2^N. With an ordered list, it can also be as bad as 2^N when the arcs have negative lengths. See Problems 22 to 24 for details.

The sizes of these bounds stem from the fact that a node can get inserted into L (in step 4 of label correction), deleted (in step 2), and inserted later. Insertions after the first are called *corrections*; Corrections mean added executions of (2-9). Intuitively, one would expect label correction to work best on sparse networks because very few corrections would be needed. In empirical tests on sparse networks with nonnegative arc lengths, label correction is competitive with reaching with heaps, provided that L is kept as a FIFO list. When L is kept as a LIFO list, label correction is inferior.

Let m be the length of the shortest arc in the network. A variant of reaching is based on the property isolated in the following theorem.

THEOREM 2. Suppose that $m > 0$. Interrupt reaching (on page 20) after any execution of step 2. With i (still) denoting the node selected in step 2, one has $v_j = f_j$ for every node j that satisfies $v_j \leq v_i + m$.

> *Remark:* Of course, the node i selected in step 2 has $v_i = f_i$. The point of Theorem 2 is that any node j whose label is within m of v_i has $v_j = f_j$.

Proof. Consider a node j having $v_j \leq v_i + m$. Aiming for a contradiction, we suppose that $v_j > f_j$. Then some subsequent execution of (2-9) will cause a decrease in v_j; that is, there exists a node k whose label was temporary when reaching was restarted and has $v_j > f_k + t_{kj}$. Since nodes are declared permanent in nondecreasing f-value sequence, $f_k \geq f_i = v_i$. So $v_j > f_k + t_{kj} \geq f_k + m \geq v_i + m$. This contradicts $v_j \leq v_i + m$, which completes the proof. ∎

We use Theorem 2 to put the nodes having temporary labels into "buckets." Suppose that $m > 0$, and let *bucket p* contain every node j whose label v_j is temporary and falls in the interval

$$(2\text{-}10) \qquad pm \leq v_j < (p + 1)m$$

These are buckets of *width m*. Consider:

Reaching with buckets of width m ($m > 0$):

1. Set $v_1 = 0$ and $v_j = \infty$ for all $j \neq 1$. (Initially, bucket 0 contains node 1, and all other buckets are empty.)
2. Stop if all buckets are empty. Else, find the lowest-numbered non-empty bucket, p^*.
3. Execute (2-9) for every arc (i, j) emanating from every node i in bucket p^*, putting node j in the appropriate bucket whenever its v-value is decreased. Then empty bucket p^*. Then go to step 2.

COROLLARY 1. Suppose that $m > 0$. Then reaching with buckets of width m terminates with $v_j = f_j$ for each j.

Proof. Reaching with buckets of width m executes (2-9) once for each arc, and Theorem 2 assures that $v_i = f_i$ when (2-9) is executed for arc (i, j). Hence, this computation terminates with a solution to (2-7); (i.e., with $v = f$). ∎

Let M be the length of the longest of the finite-length arcs; that is,

$$(2\text{-}11) \qquad M = \max \{t_{ij} \mid t_{ij} < \infty\}$$

The following exercise asks you to verify that reaching with buckets of width m requires $1 + \lceil M/m \rceil$ buckets, where $\lceil z \rceil$ denotes the smallest integer y satisfying $y \geq z$.

Exercise 6. (a) Show that at the start of any execution of step 2 of reaching with buckets of width m, every finite label v_j satisfies $v_j < (m + 1)p^* + M$.
[Hint: Justify $v_j = f_k + t_{kj}$ and $f_k < (m + 1)p^*$.]
(b) Show that a system of $1 + \lceil M/m \rceil$ buckets suffices.
[Hint: Recycle the emptied ones.]

More buckets means more movements from bucket to bucket. One would expect reaching with buckets to work best when the number of buckets is small. Empirically, it is competitive with the other methods for dense and sparse networks, provided that the number of buckets is below 1000 or 2000. In the worst case, the work needed to execute reaching with buckets of width m is proportional to the largest of the three quantities, M/m, nN, and NM/mw, where w is the number of bits per computer word. (The significance of w is indicated in Problem 27.) This bound seems cumbersome, but it sometimes offers better worst-case performance than do the other methods.

This section describes six methods for finding a shortest-path tree in a directed network. Properties of each method are recapitulated in Table 2-2.

TABLE 2-2. Six ways to compute shortest-path trees

Method of Computation	Work Bounds		Requires
	Best Case	Worst Case	
Reaching (basic)	N^2	N^2	Nonnegative arc lengths
Reaching with heaps and lists	nN	$nN \log_2 N$	Nonnegative arc lengths
Label correction with L kept as a:			
FIFO list	nN	nN^2	
LIFO list	nN	2^{N-1}	No cycles* of negative length
Ordered list	N^2	2^{N-1}	
Reaching with buckets	nN	$\max\begin{cases} nN \\ M/m \\ NM/mw \end{cases}$	Positive arc lengths

*Label correction can be adapted to detect cycles of negative length.

Listed with each method are its work bound in the most favorable case, its work bound in the least favorable case, and the class of networks on which it works. A worst-case work bound of N^2 means that the number of computer operations needed to find the shortest-path tree is bounded from above by some constant times N^2.

Several of the methods considered here exhibit best-case performance that is far superior to their worst-case performance. One would like to know how such methods perform "on average." Glimpses have been provided here of the findings obtained by running methods on representative networks. No theoretical studies are provided here of average performance, partly because it is so hard to average over representative networks.

The study of shortest-path problems in cyclic networks is extended in the problems at the end of this chapter. Problems 13 to 15 survey the literature on the problem of finding shortest paths between all pairs of nodes. Problems 16 to 18 survey the literature on kth shortest paths. Problem 25 notes that the bucket width can be increased from m to $m + 1$ when all arc lengths are positive integers. Problem 28 applies reaching with buckets to the so-called cyclic group knapsack network that arises in integer programming.

BIBLIOGRAPHIC NOTES

A large literature exists on shortest-route problems in networks whose arc lengths are nonnegative. Dijkstra (1959) is usually credited with adapting reaching to these models, although Pollack and Wiebson (1960) credit Minty. No one seems to get credit for keeping T as a heap. Early label-correcting algorithms include those of Ford (1956), Dantzig (1957), Moore (1957), and Bellman (1958). A precursor to buckets is credited to Loubal by Hitchner (1968). Dreyfus (1969) and Gilsinn and Witzgall (1973) provide excellent surveys of shortest-route methods for dense and sparse networks, respectively. Denardo and Fox (1979a, 1979b) contain extensive discussions of buckets and of other ways to accelerate reaching by exploiting problem structure. They also update the earlier surveys.

PROBLEMS

1. A company that makes turbines has one of them on hand at the beginning of the current month. Orders for this month and for the next three are for 2, 1, 1, and 0 turbines, respectively. The company wishes to have one turbine on hand at the beginning of the fifth month, which means that a total of four turbines must be manufactured during the next 4 months. Orders for a particular month may be filled from that month's production or from inventory. The problem is to find the production schedule that satisfied demand and minimizes total cost over the 4-month period. The cost of producing 0, 1, or 2 turbines in a given month is 10, 17, and 20, respectively. The cost of having 0, 1, or 2 turbines in inventory at the start of a month is 0, 3, and 7, respectively.

   Draw a network whose shortest path corresponds to the best production schedule. Find the best production schedule by recursive fixing.

2. There follows a 4 × 5 array of numbers. Suppose that an individual movement in the array is allowed rightward one column or downward one row, but not both simultaneously. Suppose that it is desired to move from the upper left corner to the lower right corner of the array by a sequence of

such movements and in a way that minimizes the sum of the integers encountered. (Seven movements are required, four to the right and three down.)

$$
\begin{array}{ccccc}
0 & 4 & 3 & 6 & 4 \\
7 & 8 & 6 & 8 & 8 \\
2 & 3 & 1 & 8 & 7 \\
6 & 2 & 9 & 3 & 0
\end{array}
$$

a. Set this up as a shortest-route problem through a directed acyclic network.

b. Compute a 4×5 array whose entries are the lengths of the shortest paths from the nodes they depict to the lower right corner of the array. Find the tree of shortest paths to the node in the lower right corner.

3. Production capacity is 3 units a month. The unit production costs are $3, $5, and $3 in months 1, 2, and 3, respectively. In addition, the cost of changing the production level from the preceding month is $2 per unit change. Inventory at the beginning of month 1 is 1 unit, and 1 unit was produced in the preceding month. Demand in months 1, 2, and 3 is for 2, 4, and 1 units, respectively. Demand must be met by production in the current month or in preceding months. Inventory carrying costs are negligible. Find a minimum-cost production plan.   27
[Hint: One approach is to establish a node for each triplet (n, i, k), where n denotes the month, i denotes the start-of-month inventory, and k denotes the preceding month's production.]

4. A construction company wishes to apply the critical path method to the scheduling of a project. The activities, their completion times, and their predecessors are as given below.

Activity	a	b	c	d	e	f
Completion Time	3	6	8	4	5	2
Predecessor(s)	—	—	a	a, b	c, d	b, c

This means, for instance, that activity d takes 4 weeks and can be started as soon as activities a and b are completed.

In a network formulation of this (or any) critical path problem, a node depicts the completion of the activities it specifies. To obtain such a formulation, use this recipe. First create a list of nodes consisting of a beginning node and an ending node; an intermediate node (such as a, ab, cd, and bc) for each set of activities in the table of predecessors; an intermediate node (such as b and c) for each activity that appears multiple times in intermediate nodes but only in combination. Then create one arc per activity, and connect them properly. For instance, the arc corresponding to activity a emanates from the beginning node and terminates at node a;

the arc for activity d emanates from node ab and terminates at node cd; and so on. Finally, add arcs of zero length, such as (a, ab), where required for logical consistency.

    a. Apply this recipe to the construction problem.

    b. Find the critical path and, for each activity, its latest allowable starting time and earliest feasible completion time.

5. Write down a linear program whose solution specifies the shortest path from node 1 to node 9 in Figure 2-3.

[**Hint:** Maximize f_1 subject to constraints suggested by (2-1) and (2-2).]

6. The following is a standard problem in the calculus of variations. With $g(\cdot, \cdot, \cdot)$ as a known function of its three arguments, find a continuous function $f(x)$ that minimizes the integral

$$\int_{x=1}^{3} g[x, f(x), f'(x)] \, dx$$

subject to the constraints $f(1) = 0$ and $f(3) = 2$. Assume that a minimizing function $f^*(x)$ exists. Draw a picture of it and then discover a version of the principle of optimality that it satisfies.

7. Joe Cool is contemplating another transcontinental migration, say, from Cambridge to Berkeley. He is wondering whether his aging auto, which tends to overheat, is up to it. He figures that altitude will give his car a sterner trial than distance. He has collected the information in Figure 2-3, where node 1 is Cambridge, node 9 is Berkeley, and the arc lengths are the altitudes (in thousands of feet) of the roads they represent. Joe wants to find a route to Berkeley whose maximal altitude is minimal.

    a. Show how to solve Joe's problem by dynamic programming.

    b. Prove or disprove: The first two versions of the principle of optimality are equivalent.

8. In many computers, arithmetic and logic operations are executed in extremely fast registers, and information is transferred as needed between these registers and the computer's main memory. Each transfer (in either direction) is called an *access*, and the execution time of an algorithm is influenced by the number of accesses it uses.

    a. How many accesses are needed to solve (2-5) by recursive fixing, assuming that the table of v-values is kept in main memory?

    b. How many accesses are needed to solve (2-6) by reaching under the same circumstance?

Note: Problems 9 to 12 are instances in which "functional" equations appear in contexts other than dynamic programming. Supplement 1 also uses functional equations; see equation (S1-3) and its Problems 6 and 7.

9. (*Summing series recursively*) Let $f(\alpha)$ denote the sum of the series $(1, \alpha, \alpha^2, \ldots)$, where $|\alpha| < 1$. It is well known that $f(\alpha) = 1/(1 - \alpha)$, but we now provide a particularly simple proof: the sum of this series equals 1 plus the sum of the series $(0, \alpha, \alpha^2, \ldots)$; the latter is α times the sum of $(1, \alpha, \alpha^2, \ldots)$, which is $f(\alpha)$, so

$$f(\alpha) = 1 + \alpha f(\alpha)$$

Now let $g(\alpha)$ be the sum of the series $(1, 2\alpha, 3\alpha^2, \ldots)$, where $|\alpha| < 1$. This series is the sum of $(1, \alpha, \alpha^2, \ldots)$ and $(0, \alpha, 2\alpha^2, 3\alpha^3, \ldots)$. Use this observation to justify

$$g(\alpha) = f(\alpha) + \alpha g(\alpha)$$

Then solve for $g(\alpha)$.

10. (*n choose k*). Let $f(n, k)$ denote the number of distinct subsets of a set having n elements, with (naturally) $f(n, k) = 0$ for $k > n$.

   a. Show that

$$f(n, k) = f(n - 1, k) + f(n, k - 1)$$

[Hint: $f(n, k - 1)$ is the number of subsets that *do* contain a particular element.]

   b. Use part a to compute $f(n, k)$ for all n and k between 0 and 5. This is a systematic way of evaluating $n!/k!(n - k)!$.

11. A pair of dice is rolled repeatedly. The probability that a "3" occurs on any given roll is $\frac{2}{36}$, and the probability that a "7" occurs on any given roll is $\frac{6}{36}$. Let $f(3)$ denote the probability that a "3" occurs before a "7." Justify and then solve

$$f(3) = \tfrac{2}{36} + (1 - \tfrac{2}{36} - \tfrac{6}{36})f(3)$$

12. (*Repeating decimals*) To express the repeating decimal $0.121212\ldots$ as a fraction, write $p/q = 0.12 + 100\,p/q$. Solve for integers p and q. Use this method to write $0.124124\ldots$ as a fraction.

Note: Problems 13 to 26 relate to the starred section on shortest routes in networks that can have cycles.

13. [Warshall (1962)]. Consider a network with nodes $1, \ldots, N$ and arc set B. Set $t_{ij} = 1$ if (i, j) is in B; otherwise, $t_{ij} = 0$. Let f_{ij} equal 1 (respectively 0) if a path exists (does not exist) from node i to node j. Show that $f_{ij}^N \equiv f_{ij}$, where $f_{ij}^0 \equiv t_{ij}$ and

$$f_{ij}^{n+1} = \max \begin{cases} f_{ij}^n \\ f_{in}^n \cdot f_{nj}^n \end{cases}$$

[Hint: Node n is inserted, whenever helpful, in the middle of a path.]

Remark: A naive implementation of this method entails work proportional to N^3; Fox and Landi (1968) give a different method whose work is proportional to N^2.

14. [Floyd (1962)]. Consider a network with nodes $1, \ldots, N$. Let t_{ij} denote the length of arc (i, j), and assume that the network has no cycles of negative length. Let f_{ij} denote the length of the shortest path from node i to node j. Show that $f_{ij}^{N+1} \equiv f_{ij}$, where $f_{ij}^1 \equiv t_{ij}$ and (as in Problem 13)

$$f_{ij}^{n+1} = \min \begin{cases} f_{ij}^n \\ f_{in}^n + f_{nj}^n \end{cases}$$

15. Alter the recursion in Problem 14 so as to detect whether a cycle exists whose length is negative.

> *Remark:* Problem 14 contains what seems to be an elegant method for finding the shortest paths between all pairs of nodes. Surprisingly, repeated use of reaching is faster. The effectiveness of algorithms for computing shortest paths between all pairs of nodes depend on whether the network is sparse or dense. For dense networks, the fastest extant methods are that of Hoffman and Winograd (1972), with respect to longest running time, and that of Spira (1973), with respect to average running time. Spira sorts the arcs by length first. For large sparse networks, repeated use of reaching with buckets may be faster still.

> *Note:* The next three problems reflect papers of Hoffman and Pavley (1959), Dreyfus (1969), and Fox (1973a) on the computation of second, third, fourth, . . ., shortest paths through directed networks having N nodes. Arcs having negative lengths are allowed here, but all cycles must have positive length. In particular, cycles consisting entirely of zero-length arcs are excluded. A common system of notation is now established. For $j \neq 1$, let $f(j, k)$ denote the length of the kth shortest path from node 1 to node j. Let $f(1, 1) = 0$ and, for $k \geq 1$, let $f(1, k + 1)$ be the length of the kth shortest cycle containing node 1. Initialize matrix $n(i, j)$ by $n(i, j) \equiv 1$. Note, as concerning shortest paths, that

> (*) $$f(j, 1) = \min_i \{f[i, n(i, j)] + t_{ij}\}, \quad j \neq 1$$

16. (Second shortest path) For each $j \neq 1$, let $n(i, j) \leftarrow [n(i, j) + 1]$ for exactly one i attaining the minimum in (*). Break ties arbitrarily.

    a. Argue that, even for $j = 1$,

(**) $$f(j, 2) = \min_i \{f[i, n(i, j)] + t_{ij}\}$$

    b. Exactly $N - 1$ arcs (i, j) will have $n(i, j) = 2$, and none of these has $j = 1$. Specify a sequence in which (**) can be evaluated.

17. (kth shortest paths) The general situation becomes clear when the third-shortest paths are computed. For each j, this time *including* $j = 1$, set $n(i, j) \leftarrow [n(i, j) + 1]$ for exactly one i attaining the minimum in (**). Break ties arbitrarily.

    a. Argue that, for each j,

(***) $$f(j, 3) = \min_i \{f[i, n(i, j)] + t_{ij}\}$$

[Hint: If, for instance, $n(i, j) = 3$, why should we append arc (i, j) to the end of the third shortest path to node i?]

    b. When evaluating (***), one must account for the possibility that $n(i, j) = 3$. Why? Can (***) be evaluated in the same sequence as was (**)?

    c. Specify a method for computing kth shortest paths. How might heaps be used to accelerate computation?

[Hint: Compare the right-hand sides of (*) and (**) and (***).]

18. Use Problems 16 and 17 to find the first, second, third, and fourth shortest paths in the network having three nodes, with $t_{ii} = \infty$ for $i = 1, 2, 3$, $1 = t_{12} = t_{21} = t_{23} = t_{32}$, and $2 = t_{13} = t_{31}$.

> *Remark:* This method takes care of ties automatically; no special bookkeeping is needed.

19. (*Managing list T in reaching*). When reaching is applied to networks having nonnegative arc lengths, one can keep T in a list of consecutive memory cells, r through N. To "delete i from T," move the label in cell r to i's cell, and increment r. Flowchart a computer code that implements this. [This way of managing T was first noted by Yen (1970).]

20. (*Label correction*) Suppose that the network has no cycle whose length is negative. Node i enters L when v_i is decreased, which happens when a shorter path is found from node 1 to node i.

   a. Show that label correction terminates finitely.

[Hint: Can node i enter L infinitely often?]

   b. Show that label correction terminates with $v_j = f_j$ for each j.

[Hint: Suppose that $v_j > f_j = f_i + t_{ij}$, and wonder about v_i.]

21. (*Detecting negative cycles*). One way to detect a cycle whose length is negative is as follows. Augment label correction to record for each node j the node $i = P(j)$ whose arc (i, j) caused the most recent reduction of v_j. Then backtrack either each time a label is reduced or intermittently, say every tenth time a label is reduced. If a cycle exists that has negative length, a node will (eventually) be repeated when backtracking. Backtracking takes time. An alternative is to augment label correcting to record for each node j the number $N(j)$ of arcs in the path attaining v_j. To do this, replace $N(j)$ by $N(i) + 1$ if executing (2-9) for arc (i, j) causes a reduction of v_j. What property of $N(\cdot)$ determines whether or not the network has a negative cycle?

22. (*Label correction with FIFO list*). Suppose that the network has no cycle whose length is negative. Consider label correction with L kept as a FIFO (first-in, first-out) list. Suppose that a shortest path from node 1 to node i has r arcs. Show that $v_i = f_i$ after r sweeps from head to tail of list L. Show that the total work is at most AN, where A is the number of arcs in the network.

   Note: The next two problems concern a family of acyclic networks whose arcs point to lower-numbered nodes. For $r = 2, 3, \ldots$, *network r* has nodes 1 through r and arcs (i, j) of length $t(i, j)$ for $r \geq i > j \geq 1$. Network r is *solved* by finding the tree of shortest paths from node r to the lower-numbered nodes. Problem 24 is based on Johnson (1973).

23. (*Label correction with LIFO list.*) Suppose that network r (as described above) is solved by label correction with L kept as a LIFO list and, specifically, with (2-9) executed for arcs emanating from node i in decreasing-j sequence [e.g., arc (3, 2) before arc (3, 1)]. Consider the data $t(i, i - 1) = 0$ and $t(i, j) = 2^{j-1} + t(i, j + 1)$ for $j < i - 2$. (So "jumping over" node k costs 2^{k-1}.)

   a. Draw network r for the case $r = 5$, and solve it by the method given in this problem. Enumerate the sequence $(5, \ldots)$ in which nodes are removed from L.

b. Let b_r denote the sequence in which nodes are removed from L when network r is solved. Write $b_r = (r, c_r)$, and justify $b_r = (r, c_{r-1}, r - 1, c_{r-1})$ for the case $r = 5$.

c. *Assume* that the preceding recursion holds for $r \geq 3$, with $b_2 = (2, 1)$. (It does.) Show that solving network r entails 2^{r-1} removals of nodes from L and $2^{r-1} - 1$ executions of (2-9). (Hence, the work is exponential in r.)

24. (*Label correction with ordered list: case of negative arc lengths.*) Suppose that network r, as described in the note preceding Problem 23, is solved by label correction with L kept as an ordered list. Consider the data $t_{ij} = -(2^{i-2} + i - j)$. Redo parts a, b, and c of Problem 23.

25. (*Reaching with buckets and integer data.*) Suppose that all arc lengths are positive integers. Adapt Corollary 1 to verify that reaching works when the buckets have width $m + 1$.

26. Ignore the arrows in Figure 2-3, thereby allowing arcs to be traversed in either direction. Use reaching with buckets of width 2 (this is justified in Problem 25) to find the tree of shortest paths from node 4 to all other nodes. How many empty buckets were encountered during computation?

27. (*Work bounds*). The *work* required by an algorithm is the number of operations required by its computer code, including additions, multiplications, memory accesses, pointer adjustments, and so on. A statement that the work is *at most* $O(Z)$ means that some constant ξ exists such that the number of computer operations is at most ξZ, for all Z. Show that the work bound for reaching with buckets of width $m > 0$ is $O(\max \{x, nN, Nx/w\})$, where x is the number of buckets needed (see Exercise 6), where w is the number of bits per computer word, and where nonempty bucket information is bit-packed into $\lceil x/w \rceil$ computer words. Assume that the computer has an instruction which shifts a word left until its leading bit is a 1 while counting the number of bits shifted in another register.

Remark: Denardo and Fox (1979a) also give work bounds for multiechelon bucket systems.

28. (*Group knapsack network*). Set $N = \{0, 1, \ldots, n - 1\}$ and $K = \{1, 2, \ldots, k\}$. Associate with each element j of K the nonnegative number c_j and the nonnegative integer α_j. For integers β and α, let $\beta \oplus \alpha = (\beta + \alpha)$ *modulo* N and $\beta \ominus \alpha = (\beta - \alpha)$ *modulo* N. A (slightly simplified) *group knapsack network* has N as its set of nodes and has arc $(\beta, \beta \oplus \alpha_j)$ of length c_j for each β in N and each j in K. (This network has n nodes, and k arcs emanate from each node.) Let $f(\beta)$ denote the length of a shortest path from node 0 to node β. With $f(0) = 0$, one gets the functional equation

$$f(\beta) = \min_{1 \leq j \leq k} \{f(\beta \ominus \alpha_j) + c_j\}, \qquad \beta \neq 0$$

Identify optimizers as follows: with $q(0) = 1$, let

$$q(\beta) = \max \{j \,|\, f(\beta) = f(\beta \ominus \alpha_j) + c_j\}, \qquad \beta \neq 0$$

a. Show that $q(\beta) \geq q[\beta \ominus \alpha_{q(\beta)}]$ for every node $\beta \neq 0$.

[Hint: Subpaths of shortest paths are shortest paths.]

b. Suppose that $0 < m = c_1 \leq c_2 \leq \ldots \leq c_k$. Suppose that reaching with buckets of width m is implemented and, at step 2, that bucket $p^* + 1$ is empty [i.e., no node β has $(p^* + 1)m \leq f(\beta) < (p^* + 2)m$]. Show that every node β that has $f(\beta) > (p^* + 1)m$ also has $q(\beta) \geq 2$.

[Hint: If not, one can pick β so that (i) $(p^* + 1)m \leq f(\beta)$, (ii) $q(\beta) = 1$, and (iii) $f(\beta)$ is minimal among the nodes satisfying (i) and (ii).]

c. Argue that the conditions identified in part b let one ignore all arcs of the type $(\beta, \beta \oplus \alpha_1)$ and increase the bucket width m from c_1 to $\min \{c_j: c_j > c_1\}$.

d. Design a bucket-widening scheme that calculates $f(\beta)$ for each β with work at most $O(kn)$.

   Note: $O(kn)$ is the work bound that would obtain for an unstructured acyclic network having the same number kn of arcs as does the group knapsack network.

   Remark: For a more comprehensive study of group knapsack networks and their use in integer programming, see Denardo and Fox (1979b), from which Problem 28 is adapted, and Denardo and Fox (1980).

3

ALLOCATION,

MARGINAL ANALYSIS,

AND LAGRANGE MULTIPLIERS

Titles above in bold type concern advanced
or specialized subjects. These sections can
be omitted without loss of continuity.

RESOURCE ALLOCATION

One of the central topics of operations research is the allocation of scarce resources. Linear allocation models of vast dimensions are routinely and quickly solved by linear programming. Nonlinear models of much more modest dimensions can be analyzed, some by dynamic programming.

You might limit your first reading of this chapter to the unstarred sections, which concern the analysis of a family of simple allocation models. These models are themselves interesting, and they illustrate the concept of a state, which is fundamental to dynamic programming. The starred sections concern efficient computation, which determines the range of applicability of these models.

ALLOCATING ONE TYPE OF RESOURCE

The simplifying feature of the first model we shall consider is that it contains only one resource. Exactly K units of that resource are available, where K is an integer. The resource can be allocated to the production of various amounts of N different commodities, which are numbered 1 through N. The problem is to do so most profitably. Commodities can only be produced in integer quantities. Producing x_n units of commodity n consumes $c_n(x_n)$ units of the resource and yields profit $p_n(x_n)$. The decision maker can leave some units of the resource unallocated, but he or she cannot produce more than B units of any single commodity. The problem of maximizing profit is expressed as the following mathematical program.

Program 1. Maximize $\{p_1(x_1) + p_2(x_2) + \ldots + p_N(x_N)\}$, subject to the constraints

(3-1) $$c_1(x_1) + c_2(x_2) + \ldots + c_N(x_N) \leq K$$

(3-2) $$0 \leq x_n \leq B, \quad x_n \text{ integer}, \quad n = 1, \ldots, N$$

The objective in Program 1 is total profit. Constraint (3-1) records the fact that resources consumed cannot exceed resources available. Constraint (3-2) reflects the fact that only certain integer quantities of each commodity can be produced. To simplify the presentation, assume that consumption occurs in integer quantities and that producing nothing consumes nothing [i.e., $c_n(x_n)$ is nonnegative and integer-valued, with $c_n(0) = 0$].

Time does not play a central role here. Yet, imagining that the N different commodities are produced one after another converts the allocation model to a sequential decision process. Suppose that commodity N is produced first, then commodity $N - 1$, and so on, with commodity 1 last. Suppose that various quantities of commodities $n + 1$ through N have already been produced. A certain number of units, say y, of the resource remains. Since each production activity consumes the resource in integer quantities, y must be some integer satisfying $0 \leq y \leq K$. These y units are to be allocated to the production of commodities 1 through n. Let

$f(n, y) =$ the maximum total profit that can be obtained from allocating y units of the resource to production of commodities 1 through n.

By definition, $f(N, K)$ is the maximum profit for the allocation problem in Program 1. This problem has just been *embedded* in a family of $N(K + 1)$ related optimization problems, namely the determination of $f(n, y)$ for $n = 1, 2, \ldots, N$ and $y = 0, 1, \ldots, K$. Only finitely many production schedules satisfy the constraints of Program 1; hence, $f(n, y)$ is attained for each n and y.

A functional equation is now developed. With y units of the resource on hand, consider producing x units of commodity n, where

(3-3) $$c_n(x) \leq y$$

Constraint (3-3) requires that consumption of the resource not exceed its availability. Production quantity x is called *feasible* for (n, y) if x is an integer between 0 and B (inclusive) and if (3-3) holds.

When n equals 1, feasibility is the only consideration, and

(3-4) $$f(1, y) = \max_{x \text{ feasible}} \{p_1(x)\}$$

Now suppose that n is greater than 1, so that at least two different commodities remain to be produced. The profit obtained by producing x units of commodity n and allocating the remaining $y - c_n(x)$ units optimally to production of commodities 1 through $n - 1$ is, by definition,

$$p_n(x) + f[n - 1, y - c_n(x)]$$

$f(n, y)$

In other words, the foregoing is the largest profit one can obtain by allocating y units of the resource to the production of commodities 1 through n, *provided* that one produces exactly x units of commodity n. Maximizing this expression over all feasible x removes the proviso and, by definition, attains $f(n, y)$.

(3-5) $$f(n, y) = \max_{x \text{ feasible}} \{p_n(x) + f[n - 1, y - c_n(x)]\}$$

You should recognize (3-4) and (3-5) as a *functional equation* of dynamic programming; (3-5) has f on both sides of the equality sign, and (3-4) and (3-5) interlock the solutions of a set of optimization problems.

To apply recursive fixing, note that (3-5) specifies $f(n, y)$ in terms of the numbers $f(n - 1, z)$ for various values of z. Also, (3-4) specifies $f(1, y)$ in terms of the basic data of the model. Consequently, *recursive fixing* amounts to the following procedure. First, use (3-4) to compute $f(1, y)$ for each integer y between 0 and K; then use (3-5) to compute $f(2, y)$ for each integer y; then use (3-5) to compute $f(3, y)$ for each y; and so on, ending with computation of $f(N, K)$.

Exercise 1. Suppose that all resources must be consumed, so that equality holds in (3-1). How does this change the functional equation?

Exercise 2. Sketch a network for which $f(n, y)$ is the longest path from node (n, y) to any ending node. Write down a recursive fixing routine such as those in Chapter 2 that computes $f(n, y)$ for each (n, y). Include a procedure for recording an optimal policy $x(\cdot, \cdot)$ while computing $f(\cdot, \cdot)$. [An optimal policy specifies for each pair (n, y) an integer $x(n, y)$ that maximizes the right-hand side of the functional equation.] Write down a routine that uses $x(\cdot, \cdot)$ to backtrack a production plan that attains $f(N, K)$.

STATES

States play a key role in dynamic programming, and the term "state" is used throughout this volume in a manner that preserves the following properties:

> *Properties of states:* Transition occurs from state to state, and a state is a summary of the prior history of the process that is sufficiently detailed to enable evaluation of current alternatives.

We illustrate. For the network model of Chapter 2, think of the nodes as states. Think of the arcs emanating from a node as depicting the various decisions available there. With this interpretation, transition occurs from state to state, and the states are the points at which decisions are made. The displayed statement boils down to the observation that arcs emanating from a node have nothing to do with how that node was reached. For instance, the length of arc (i, j) is independent of the path that led to node i.

For the allocation model, take the pair (n, y) as a state. State (n, y) depicts the situation of allocating y units of the resource to the production of commodities 1 through n. The information contained in state (n, y) suffices to determine whether production quantity x is feasible and, if it is, the state to which transition occurs. In short, enough information is contained in the pair (n, y) to enumerate and evaluate current options.

The set of states in a model is called its *state space*.

Exercise 3. In the allocation model, the singletons n and y do not suffice as states. Why? Also, the triplet (n, x_{n+1}, y) does suffice as a state, where x_{n+1} is the production of commodity $n + 1$. Why? Do you have a preference?

The term "state" has comparable meanings in other fields. In Markov chains, the law of motion depends only on the current state, and not on states visited previously. In physics, chemistry, and control theory, the state of the system is usually a summary of its past behavior that is sufficient to predict evolution and response to stimuli. In all these cases, transition occurs from state to state. But you should be informed and forewarned that the term "state" is used in several ways in operations research. Many writers on dynamic programming separate the pair (n, y) into a *stage n* and a *state y*. We call the pair (n, y) a state, as only this usage preserves the displayed interpretation. Also, in queueing theory, the number of customers in the system is often taken as the state, even though it is insufficient to determine the law of motion in all but the simplest case.

The elements of a state are called *state variables*. For instance, the allocation model has two state variables, n and y.

Several facts contribute to accord states a central role in dynamic programming. A sequential decision process evolves from state to state. States are the points at which decisions are made. The functional equation and the principle of optimality link the optimal return to the starting state.

A further reason why states are so important is the following: Formulating an optimization problem for solution by dynamic programming amounts to a quest for a state space. Finding a state space can be difficult, especially for beginners. The properties that states always satisfy serve as guideposts and ease this search. For instance, the key to converting the allocation model to a dynamic programming problem is the observation that when things are produced in fixed sequence, it is only necessary to remember the item n now being produced and the number y of units of resource that remain, as n and y suffice to evaluate the current options.

AN EXAMPLE

When formulating an optimization problem for solution by dynamic programming, look for the states. Remember that a state must be a summary of what went before that is sufficiently detailed to allow the evaluation of current alternatives. To illustrate, we now work a whimsical example. Try doing it yourself before reading the solution. And remember that the key is to define the states properly.

Example.   Trout Rader is planning on poaching in the national preserve, which contains K lakes. A careful planner, he has assembled the following data.

$y_i =$ the number of trout he expects to catch in 1 hour's fishing at lake i

$t_{ij} =$ the travel time (an integer) in hours from lake i to lake j

Of course, travel times satisfy the triangle inequality, $t_{ik} \le t_{ij} + t_{jk}$. Trout recognizes that there is some risk of capture by the game warden. To minimize this risk, he has decided to fish for only 1 hour in each lake he visits. This offers him some protection against the possibility that a passerby finds the game warden and brings him back. Trout allows himself the option of returning to a lake after having fished in another, as the game warden will likely have left. Trout has just entered the preserve at lake 1. It is 12 hours until suppertime. How shall he poach so as to maximize his catch until then?

   Solution: What constitutes a state? It is clear that Trout must keep track of the number of hours that remain until suppertime and of the lake he is at. He need not keep track of the lakes he has visited previously, as he allows himself the option of revisiting a lake after leaving it. Nor need he keep track of the number of trout caught previously, as this will not affect his future catch. The formulation has two complicating details; it may be optimal to leave lake 1 immediately, and it may be optimal to fish for fewer than 12 hours.

   Let us try the pair (n, i) as a state, this denoting the situation of leaving lake i with n hours left before suppertime. That is, i and n are the state variables. Define $f(n, i)$ as the largest number of fish Trout can catch in n hours given that he is leaving lake i; include in $f(n, i)$ the y_i fish he just caught. Trout's decision is where to fish next. If he chooses lake j, he leaves there $t_{ij} + 1$ hours later. This justifies the functional equation[1]

$$\textbf{(3-6)} \qquad f(n, i) = \begin{cases} y_i + \max\limits_{j\,|\,j \neq i} \{f(n - t_{ij} - 1, j)\}, & n \ge 0 \\ 0, & n < 0 \end{cases}$$

If Trout fishes for 1 hour in lake 1, he gets a total of $f(11, 1)$ fish. If he departs immediately, he gets $f(12, 1) - y_1$ fish.

   The aforegoing is only one of several correct formulations. If yours differs, the difference may stem from a slightly different definition of a state, which leads to an expression that differs slightly from (3-6).

---

[1] The vertical line stands for "such that" (e.g., the expression in braces is maximized over all j other than i).

The most common error in dynamic programming formulations is to account improperly for the end of the planning horizon. It is remarkably easy to get these boundary conditions wrong! The simplest check for correct boundary conditions is to work a small example. If your equation differs from (3-6), you might test it on the data in Problem 1.

Exercise 4. Write a functional equation for which state (n, i) is interpreted as arriving at lake i with n hours left.

Exercise 5. Suppose that Trout's wife is camping at lake 1, and that she wants him to return there so that his catch may be poached for supper. Alter (3-6) to reflect this.

A KNAPSACK MODEL

Let us consider the special case of the allocation model in which the production and consumption functions are both linear. This version has a second and simpler formulation as a sequential decision process.

Interpret the integer K as the volume of a knapsack. Each unit of commodity n now occupies volume c_n (a positive integer) in the knapsack and yields profit p_n. The optimization problem is to find the most profitable way to pack the sack. To simplify the model's network formulation, we concoct commodity 0 in a way that has the same effect as allowing the sack to remain partially empty. Each unit of commodity 0 occupies 1 unit of volume ($c_0 = 1$) and produces no profit ($p_0 = 0$). In mathematical terms, the problem of how best to pack the sack is

Program 2: Maximize $\{p_0 x_0 + \ldots + p_N x_N\}$, subject to the constraints

$$c_0 x_0 + \ldots + c_N x_N = K$$
$$x_n \geq 0, \quad x_n \text{ integer}, \quad n = 0, \ldots, N$$

Hikers will recognize the knapsack problem as a gross oversimplification of their packing problem. It is, however, a prototypical integer program.

The constraint that x_n be an integer is crucial, because one would otherwise fill the sack with the commodity having largest ratio p_n/c_n of unit profit to unit cost. It is assumed here that no two commodities have the same unit volume. This entails no loss of generality, as the less profitable commodity would never be packed.

Program 2 is the special case of Program 1 in which all the profit functions and consumption functions are linear. The functional equation that applies to the general case also works here, but we shall see that it is unnecessarily complex.

Imagine that some items have been put in the sack and that j units of volume remain. Since the production and consumption functions are linear, the problem of packing the remainder of the sack most profitably is independent of what was put into it previously. Hence, j is a state. For $j = 0, 1, \ldots, K$, let

$f(j) = $ the maximum profit obtainable from packing a sack
whose volume is j units

Putting 1 unit of commodity n in a sack of volume j yields profit p_n and reduces the available volume to $j - c_n$. With $f(0) = 0$, one gets the functional equation

(3-7) $$f(j) = \max_{n \mid c_n \leq j} \{p_n + f(j - c_n)\}, \qquad j = 1, \ldots, K$$

If commodity 0 were omitted, this functional equation would be incorrect, unless one interpreted the maximum over an empty set as zero. Recursive fixing consists of evaluating (3-7) successively for j equal to 1 through K. The analogy between functional equations (3-7) and (3-5) suggests that the symbol y, not j, should be used to describe the state. But we will need to relate (3-7) to the material on reaching in Chapter 2, where j is used.

The knapsack model requires only one state variable, as opposed to two for the general model. The recursive fixing procedure for the knapsack model entails K states and $N + 1$ evaluations per state. The recursive fixing procedure for the general allocation model entails $N(K + 1)$ states and several decisions per state. This clearly indicates that the knapsack formulation is simpler and faster.

Exercise 6. Suppose that all profit and consumption functions are linear, except for commodity N. Can you think of an alternative to reverting completely to (3-4) and (3-5)?

MULTIPLE RESOURCES

The models considered so far concern the allocation of K units of *one* resource to the production of various quantities of N different commodities. Such models fail to encompass the situation in which production of any single commodity requires consumption of several different resources. It would be very desirable to generalize the model to multiple resources. We will soon see that this generalization entails no conceptual difficulties, but that it can create a severe computational problem.

Let us consider a model having just two resources. The second resource is modeled after the first. Specifically, assume that L units of the second resource are available. Producing x units of commodity n now consumes $d_n(x)$ units of the second resource as well as $c_n(x)$ units of the first resource. The consumption functions $c_n(x)$ and $d_n(x)$ are assumed to take nonnegative integer values, with $c_n(0) = d_n(0) = 0$ for each n. If he or she wishes, the decision maker may leave some units of either resource unallocated. The problem of allocating the two resources so as to maximize profit has the following mathematical representation.

Program 3: Maximize $\{p_1(x_1) + \ldots + p_N(x_N)\}$, subject to the constraints

$$c_1(x_1) + \ldots + c_N(x_N) \leq K$$
$$d_1(x_1) + \ldots + d_N(x_N) \leq L$$
$$0 \leq x_n \leq B, \qquad x_n \text{ integer}, \qquad n = 1, \ldots, N$$

Programs 1 and 3 differ only in that the latter has an extra constraint.

What now constitutes a state? Suppose that production quantities have been set for commodities $n + 1$ through N and that some of the resources have been consumed. There remain y units of the first resource and z units of the second resource. Allocation of these units to production of commodities 1 through n can be made without knowing the other production levels. So the triplet (n, y, z) now constitutes a state.

Conceptually, the situation is just as it was. Production level x is now called *feasible* for state (n, y, z) if x is an integer between 0 and B (inclusive) that satisfies the inequalities

$$c_n(x) \leq y, \qquad d_n(x) \leq z$$

A familiar argument justifies the following functional equation; for $n > 1$,

$$f(n, y, z) = \max_{x \text{ feasible}} \{p_n(x) + f[n - 1, y - c_n(x), z - d_n(x)]\}$$

Exercise 7. Write down an expression for $f(1, x, y)$. Justify it and the preceding functional equation.

It is now clear that introducing a second type of resource entails no conceptual difficulty. The number of states has grown, however. Originally, the states ranged over all pairs (n, y). Now they range over all triplets (n, y, z). Because of the addition of the third state variable, the number of states has grown by the factor $(L + 1)$. Recursive fixing requires one maximization per state, and the volume of computation grows in rough proportion to the number of states.

Introducing L units of the second resource *multiplies* the volume of computation by the factor of roughly $(L + 1)$. Suppose that M units of a third resource are introduced. The states become quadruplets, instead of triplets, and computation increases by the factor $M + 1$. As additional resources are introduced, additional variables must be incorporated into the definition of a state, and computation grows by multiplicative factors—that is, exponentially in the number of state variables. This rather unfortunate aspect of dynamic programming has been dubbed *the curse of dimensionality*. A large modern computer might accommodate a model having as many as 10^6 or even 10^9 states. It is easy to see that adding resources to the allocation model quickly exhausts this capacity.

This discussion points out the strength and the weakness of dynamic programming. Dynamic programming does not presume much in the way of regular behavior on the consumption and production functions. Nor does it exploit whatever regular behavior these functions might exhibit. In this sense, the curse of dimensionality is the price of generality.

SUMMARY

The most crucial concept in the preceding sections is that transitions occur from state to state, the states being summaries of the prior history of the process that are sufficiently detailed for current purposes. The preceding sections describe a useful family of allocation models. They provide added exposure to the concepts of embedding, the functional equation, the principle of optimality, and recursive fixing. The multicommodity allocation model illustrates the curse of dimensionality.

The sections that follow are starred, and they concern efficient computation. Reaching is accelerated by exploiting the special structure of the allocation models, according it an advantage over recursive fixing. Marginal analysis is shown to have advantages in speed and in flexibility when it applies, even approximately. Lagrange multipliers sometimes circumvent the curse of dimensionality, and they reveal a relationship between linear programming and dynamic programming.

REACHING*

Reaching is now applied to the knapsack and the allocation models. Both of these models have enough special structure to allow reaching to be accelerated in ways that accord it an advantage over recursive fixing.

The Knapsack Model Revisited

The ideas we shall use to accelerate reaching are of a sort that are easiest to grasp by working an example. Table 3-1 contains the data of a two-commodity (three, if commodity 0 is counted) knapsack problem. In general and in this instance, each unit of commodity n occupies c_n units of volume and yields profit p_n. Commodity 0 has $c_0 = 1$ and $p_0 = 0$, which has the same effect as letting the sack remain partially empty.

Reaching was introduced in Chapter 2 in the context of shortest-route and longest-route problems. To view the knapsack problem as a longest-route problem, proceed as follows. Create nodes $0, 1, \ldots, K$. Create arc $(i, i + c_n)$ of length p_n for all integers i and n that satisfy $0 \leq i$ and $i + c_n \leq K$.

---

* This section can be omitted without loss of continuity.

TABLE 3-1. A two-commodity knapsack problem

n	0	1	2
c_n	1	3	5
p_n	0	6	9

The quantity $f(j)$ was originally the largest profit obtainable from packing a sack having j units of volume. A network has just been constructed in such a way that $f(j)$ is the length of the longest path from node 0 to node j. If this fact seems at all mysterious, it may help to work the exercise that follows our adaptation of reaching to the knapsack problem.

Reaching procedure:

  1. Set $v(j) = 0$ for $j = 0, \ldots, K$.
  2. DO for $i = 0, \ldots, K - 1$.
    3. DO for $n = 0, \ldots, N$.
      4. Set $j \leftarrow (i + c_n)$. If $j \leq K$, set $v(j) \leftarrow \max \{v(j), v(i) + p_n\}$.

Note: The nest of DO loops executes step 4 in increasing i and, for each value of i, in increasing n.

Exercise 8. Draw the network appropriate to the data in Table 3-1, with $K = 12$. Apply reaching to this network.

Reaching, as just specified, terminates with $v(j) = f(j)$ for each j. It is now augmented to record an optimal packing plan. In addition, as computation proceeds, information gleaned about the solution to the knapsack problem is used to ignore certain of its network's arcs. These arcs get *pruned* during computation.

The value of the items in a sack is not affected by the sequence in which they are inserted. We shall contrive to restrict attention to paths emanating from node 0 whose initial arcs correspond to packing commodity N, followed by arcs corresponding to commodity $N - 1$, and so forth, ending with arcs corresponding to commodity 0. Commodities will, in this sense, be packed in *decreasing sequence*.

Reexamine functional equation (3-7) in this light, and note that, for any fixed value of j, the expression $p_n + f(j - c_n)$ is maximized by every commodity n that is included in any optimal packing of a sack whose volume is j. Consider, as an illustration, the data in Table 3-1. The expression $p_n + f(8 - c_n)$ is maximized by $n = 1$ *and* by $n = 2$, as it is optimal to include 1 unit each of commodities 1 and 2 in a sack having 8 units of volume. We shall break ties in (3-7) by defining $m(j)$ as the commodity with smallest index that it is optimal to pack in a sack having j units of volume. That is, set

$$m(j) = \min \{n \,|\, f(j) = p_n + f(j - c_n)\}, \qquad j \geq 1$$

It also proves convenient to set $m(0) = N$.

So it is optimal to place item $m(j)$ in a sack whose volume is j.

Exercise 8 (continued): For the network you were asked to draw earlier, specify $m(j)$ for each j. What packing plan in optimal? Specify a general procedure for "backtracking" an optimal packing plan from the $m(j)$'s.

Now suppose that $m(j)$ equals n, and consider $m(j - c_n)$. A set of items that it is optimal to include in a sack having volume j contains the set of items that it is optimal to pack in a sack of volume $j - c_n$. This means that $m(j)$, which is the lowest-indexed item in the larger set, is no greater than $m(j - c_n)$, which is the lowest-indexed item in the smaller set. [Can you see why we set $m(0)$ equal to N?] The preceding is the essence of a proof of the first assertion in Lemma 1.

LEMMA 1.   Consider any $j \geq 1$, and set $n = m(j)$. Then

(3-8) $m(j - c_n) \geq n$

Let $M = \max_k \{c_k\}$. If $p_1/c_1 \geq p_k/c_k$ for all k, then

(3-9) $f(j) = p_1 + f(j - c_1)$ whenever $j > (c_1 - 1)M$

> *Remark:* Expression (3-9) shows that the optimization problem disappears for large j. For $j > (c_1 - 1)M$, it has a *periodic* solution, and its period equals c_1 or divides c_1. The proof is technical; you may wish to skip over it.

Proof. Problem 10 sketches a straightforward proof of (3-8). To demonstrate (3-9), we consider any j for which $f(j) > p_1 + f(j - c_1)$, and we let $(n_0, \ldots, n_m)$ denote an optimal (longest) path from node 0 to node j (so $n_0 = 0$ and $n_m = j$). There are m arcs in this path, and the inequality $c_k \leq M$ assures that $j \leq mM$. So it suffices for (3-9) to demonstrate that $m \leq c_1 - 1$. Aiming for a contradiction, suppose that $m \geq c_1$. For $i = 0, \ldots, m$, let

$$z(i) = (n_i) \ modulo \ c_1$$

The function z maps the integers 0 through m into the integers 0 through $c_1 - 1$. As $m \geq c_1$, the first set is larger, and there must exist $i < j$ such that $z(i) = z(j)$. Consequently, $(n_j - n_i)$ is a multiple of c_1. As commodity 1 has the largest ratio of profit to space, path $(n_i, \ldots, n_j)$ must have length $(n_j - n_i)p_1/c_1$, as this corresponds to packing exactly $(n_j - n_i)/c_1$ units of commodity 1. Substitute commodity 1 for the subpath $(n_i, \ldots, n_j)$ to obtain an optimal path from node 0 to node j that includes one or more arcs corresponding to commodity 1. Hence, $f(j) = p_1 + f(j - c_1)$, which is the desired contradiction. This assures that $m \leq c_1 - 1$, which completes the proof.   ∎

Reaching is now augmented to exploit (3-8) but not (yet) (3-9). It would suffice to execute step 4 of reaching for each pair (i, n) such that, with $j = i + c_n$, one has $n = m(j)$. This seems to be a pointless observation; since $j > i$, one would need to be clairvoyant to know $m(j)$ when step 4 is executed for the pair (i, n). But (3-8) assures us that $m(j) \leq m(i)$; hence, it suffices to execute step 4 of reaching for all pairs (i, n) having $n \leq m(i)$. A version of reaching that does this is specified below.

Reaching procedure (*accelerated*):

1. Set $v(j) = 0$ and $m(j) = N$ for $j = 0, \dots, K$.
2. DO for $i = 0, \dots, K - 1$.
  3. DO for $n = 0, \dots, m(i)$.
    4. Set $j \leftarrow (i + c_n)$ and $t \leftarrow v(j)$. If $j > K$, go to step 5.
      Otherwise, set $v(j) \leftarrow \max \{t, v(i) + p_n\}$
      If $t < v(j)$, set $m(j) \leftarrow n$.
      If $t = v(i) + p_n$, set $m(j) \leftarrow \min \{m(j), n\}$.
  5. CONTINUE.

Note: The CONTINUE statement causes step 4 to be repeated with the "next" pair (i, n) [i.e., with n replaced by $(n + 1)$ when $n < m(i)$]. The above is written so as to ease its reading; you may spot repeated arithmetic and logic.

The accelerated version of reaching executes step 4 for those pairs (i, n) having $n \leq m(i)$. The estimate $v(j)$ of $f(j)$ is improved (increased) if possible. If an improvement occurs, the new estimate of $m(j)$ is recorded. If a tie occurs [i.e., if $t = v(i) + p_n$], the estimate of $m(j)$ is replaced by the smaller of itself and n.

Exercise 8 (concluded): Apply the accelerated form of reaching to the network you drew previously. Compare your results with those in Figure 3-1. How many arcs were excluded by executing step 4 for $n = 0, \dots, m(i)$?

Figure 3-1 illustrates the fact that commodities are packed in decreasing sequence. Note that Figure 3-1 contains the path $(0, 5, 8)$ but not the path $(0, 3, 8)$. The former corresponds to starting at node 0, packing 1 unit of commodity 2, which causes transition to node 5, and then packing 1 unit of com-

FIGURE 3-1. Reaching, accelerated

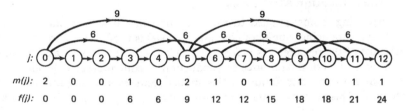

modity 1, which causes transition to node 8. Path $(0, 3, 8)$ corresponds to packing the same commodities, although not in decreasing sequence, and it was excluded. Reaching has been accelerated by *pruning* the network of some of its arcs.

Expression (3-9) has not yet been exploited. It rests on the assumption that the ratio p_k/c_k is maximized by $k = 1$. This can always be accomplished, if necessary, by relabeling the commodities. Expression (3-9) means that we need only compute $f(j)$ for $j \leq (c_1 - 1)M$, as the problem of computing $f(j)$ for larger j has then been solved. Let

(3-10) $i^* = \max\{j \,|\, m(j) > 1\}$

It is immediate from (3-9) that

(3-11) $i^* \leq (c_1 - 1)M$

Bounds like (3-11) rest on worst-case analyses and tend to be conservative. The network in Figure 3-1 has, for instance, $i^* = 5$ and $(c_1 - 1)M = 2 \cdot 5 = 10$. Problem 11 contains a different bound on the point at which the knapsack problem has become periodic. As worst-case bounds tend to be conservative, one might include either of the following techniques in a code that implements reaching. Check occasionally (say, when $i = 10, 20, \ldots$) whether all arcs corresponding to commodities 2 through N have been pruned from the network. Alternatively, maintain a list of these arcs, stopping when it becomes empty.

Exercise 9. Apply the accelerated form of reaching to the displayed data; find i^* and the point at which the network is pruned of all arcs corresponding to commodities 2 and 3.

n	0	1	2	3
c_n	1	3	4	2
p_n	0	6	7	3

Note that (3-8) is better adapted to reaching than to recursive fixing because i varies on the outer loop of reaching. When using recursive fixing, it would take almost as much effort to check whether (3-8) holds as it does to execute step 4, which may be necessary anyhow.

The Allocation Model Revisited

Reaching is now applied to the allocation model having one resource. Recall that producing x units of commodity n yields profit $p_n(x)$ and consumes $c_n(x)$ units of the resource. As before, $f(n, y)$ is the largest profit possible when allocating y units of the resource to production of commodities 1 through n.

Our network formulation of this model contains a node (n, y) for each state (n, y). This network contains one additional node, $(0, 0)$, and the network's arcs will be defined so that the length of the longest path from node $(0, 0)$ to node (n, y) is the largest profit possible when allocating y units of the resource to production of commodities 1 through n. We now call production level x_n of commodity n *feasible* for node $(n - 1, y)$ if the following conditions are satisfied:

(3-12)
$$x_n \in \{0, 1, \ldots, B\}$$

(3-13)
$$y + c_n(x_n) \leq K$$

For each node $(n - 1, y)$ and each x_n that is feasible for node $(n - 1, y)$, create an arc of length $p_n(x)$ that emanates from node $(n - 1, y)$ and terminates at node $[n, y + c_n(x)]$. With this construction, the longest path from node $(0, 0)$ to node (n, y) corresponds to the largest profit obtainable by allocating *exactly* y units of the resource to production of commodities 1 through $n - 1$. However, not all of the resource needs be consumed because (3-1) is an inequality. To allow some resource to go unused, create *vertical* arcs of length 0 that emanate from node $(n, y - 1)$ and terminate at node (n, y) for $1 \leq n \leq N$ and $1 \leq y \leq K$.

With this construction, $f(n, y)$ is the length of the longest path from node $(0, 0)$ to node (n, y). As usual, reaching makes one comparison for each arc emanating from node (n, y). For the vertical arc, it replaces $v(n, y + 1)$ with $v(n, y)$ whenever the latter is larger. For any other arc emanating from node (n, y), reaching replaces $v[n + 1, y + c_{n+1}(x)]$ by $v(n, y) + p_{n+1}(x)$ whenever the latter is larger. These comparisons are executed for all arcs emanating from node $(0, 0)$, then all from nodes $(1, y)$ in increasing y, then nodes $(2, y)$ in increasing y, and so on.

As just specified, reaching is just as fast as recursive fixing. But some arcs will shortly be eliminated. Notice that

(3-14)
$$f(n, y) \geq f(n, y - 1)$$

as the allocation attaining $f(n, y - 1)$ remains feasible when the resource level is increased to y. Consider:

LEMMA 2. Implementing the following rule cannot eliminate all longest paths from node $(0, 0)$ to any node.

> *Rule:* If $f(n, y) = f(n, y - 1)$, do not create any arcs emanating from node (n, y) except the vertical arc to node $(n, y + 1)$.

Proof. Aiming for a contradiction, assume that this rule eliminates all longest paths from node $(0, 0)$ to some node (m, w) with w *minimal*. A longest path from $(0, 0)$ to (m, w) was deleted. Consequently, this path must contain an intermediate node (n, y) for which

(3-15)
$$f(n, y) = f(n, y - 1)$$

Its subpath from (n, y) to (m, w) corresponds to certain production quantities of commodities $n + 1$ through m. These production quantities $x_{n+1}, \ldots, x_m$ yield a certain profit p, and

(3-16)
$$f(m, w) = f(n, y) + p$$

Production quantities $x_{n+1}, \ldots, x_m$ consume $w - y$ units of the resource, and they also correspond to a path of length p from node $(n, y - 1)$ to node $(m, w - 1)$. Consequently,

(3-17) $$f(m, w - 1) \geq f(n, y - 1) + p$$

Combine (3-15) to (3-17) into

$$f(m, w - 1) \geq f(m, w)$$

The above and (3-14) assure that $f(m, w) = f(m, w - 1)$. But, by hypothesis, not all optimal paths from $(0, 0)$ to $(m, w - 1)$ were deleted, and the vertical arc is never deleted. Hence, an optimal path to node (m, w) exists, establishing the desired contradiction. ■

A reaching procedure that incorporates this rule is summarized below, with the vertical arcs evaluated in step 4 and the others in step 5.

Reaching (accelerated):

    1. Set $v(0, 0) = 0$ and $v(n, y) = -\infty$ otherwise. For $x = 0, 1, \ldots, B$, set $v[1, c_1(x)] \leftarrow \max \{v[1, c_1(x)], p_1(x)\}$.

    2. DO for $n = 1, \ldots, N$.

        3. DO for $y = 0, \ldots, K$.

            4. If $y \geq 1$ and $v(n, y) \leq v(n, y - 1)$, set $v(n, y) \leftarrow v(n, y - 1)$, and go to step 7. If $n = N$, go to step 7.

            5. Otherwise, DO for $x = 0, \ldots, B$.

                6. Set $z = y + c_{n+1}(x)$. If $z \leq K$, set

$$v(n + 1, z) \longleftarrow \max \begin{cases} v(n + 1, z) \\ v(n, y) + p_{n+1}(x) \end{cases}$$

        7. CONTINUE.

This illustrates an advantage of reaching. With respect to the sequence of computation, reaching associates arcs with nodes whose longest paths have been computed, whereas recursive fixing associates arcs with nodes whose longest paths have yet to be computed. Reaching can be adapted to exploit whatever structure exists on the set of longest paths; recursive fixing cannot be. We have just exploited (3-14) in such a way. Reaching may not reduce the worst-case work bound, but it makes typical problems run faster.

Exercise 10.   What changes are needed in the network if all resources must be consumed, so that equality holds in (3-1)? Can reaching be accelerated?

Exercise 11.   Reconsider Program 3, which depicts an allocation model having two types of resources. What replaces (3-14)? What rule would you invoke to accelerate reaching?

DECREASING MARGINAL RETURN*

Consider a function g whose domain is the set of nonnegative integers. Its first forward difference $\Delta g(x)$ is defined by

$$\Delta g(x) = g(x + 1) - g(x), \qquad x = 0, 1, \ldots$$

The function g is called *concave* if its first forward difference is nonincreasing—that is, if

(3-18) $\qquad \Delta g(x) \geq \Delta g(x + 1), \qquad x = 0, 1, \ldots$

Interpret $g(x)$ as the benefit that accrues from possessing x units, so that $\Delta g(x)$ is the marginal benefit of the next unit. A function satisfying (3-18) is also said to exhibit *decreasing marginal return*, as the marginal benefit of each unit is no more than that of the last. (This term is synonymous with concavity.) Consider Program 4.

Program 4: Maximize $\{p_0(x_0) + p_1(x_1) + \ldots + p_N(x_N)\}$, subject to the constraints

$$x_0 + x_1 + \ldots + x_N = K$$
$$x_n \in \{0, 1, 2, \ldots\} \qquad \text{for } n = 0, \ldots, N$$

Assume that $p_0(x) = 0$ for all x, so that commodity 0 plays the role of a slack variable. Program 4 is studied under the assumption that each profit function has decreasing marginal return. That is, with

$$\Delta p_n(x) = p_n(x + 1) - p_n(x)$$

we shall study Program 4 under the assumption that

(3-19) $\qquad \Delta p_n(x) \geq \Delta p_n(x + 1), \qquad n = 1, \ldots, N, \quad x = 0, 1, \ldots$

Toward the end of this section, (3-19) will be relaxed.

An allocation model satisfying (3-19) can arise in several ways. Suppose, for instance, that a distributor has purchased K crates of a perishable commodity—say, bananas—and wishes to distribute them among N supermarkets as to maximize sales, which amounts to minimizing spoilage. She estimates that sending x crates to supermarket n results in sales of $p_n(x)$ crates and spoilage of the remainder, which is $x - p_n(x)$ crates. It is reasonable to assume that each additional crate she sends to supermarket n has increased chance of going unsold. That is, spoilage increases marginally (and sales decrease marginally) in x.

---

* This section concerns efficient computation and can be omitted without loss of continuity. Supplements 1 and 2 are used in minor ways here.

Dynamic programming has been espoused repeatedly as appropriate for solving this model. Equations (3-4) and (3-5) characterize its optimal solution, and recursive fixing computes it. However, we shall see that a technique called marginal analysis is both faster and more flexible. In preparation, let us call allocation $\mathbf{x} = (x_0, \ldots, x_N)$ *feasible* if it satisfies the constraints of Program 4.

Marginal analysis considers two related questions about a feasible allocation $\mathbf{x}$. Is it profitable to increase production of some commodity by 1 unit and decrease production of some other commodity by 1 unit? If so, which switch increases profit the most? Let

$$I(\mathbf{x}) = \max \{\Delta p_n(x_n)\}$$
$$D(\mathbf{x}) = \min_{m \,|\, x_m > 0} \{\Delta p_m(x_m - 1)\}$$

Note that $I(\mathbf{x})$ is the largest increment in profit for increasing any x_n by 1 and that $D(\mathbf{x})$ is the smallest reduction in profit for decreasing any x_m by 1. If $I(\mathbf{x}) > D(\mathbf{x})$, profit is increased by the switch that increases x_n by 1 and reduces x_m by 1.

Marginal analysis computes $I(\mathbf{x})$ and $D(\mathbf{x})$ and tests whether $I(\mathbf{x}) > D(\mathbf{x})$. If so, the indicated switch is executed, and the process is repeated. In other words, *marginal analysis* consists of starting with a feasible allocation $\mathbf{x}$ and repeating the following trading step until it stops.

Trading step: Stop if $I(\mathbf{x}) \leq D(\mathbf{x})$. Otherwise, increase x_n by 1 and decrease x_m by 1, where n and m satisfy

$$I(\mathbf{x}) = \Delta p_n(x_n), \qquad D(\mathbf{x}) = \Delta p_m(x_m - 1)$$

Exercise 12. For marginal analysis to be well defined, the maxima and minima must be attained for different indices; that is, $m \neq n$. Does (3-19) guarantee $m \neq n$?

Each switch increases total profit and preserves feasibility. The number of feasible allocations is finite, so marginal analysis terminates in finitely many steps. It terminates when it identifies a feasible allocation $\mathbf{x}^*$ having

(3-20) $$I(\mathbf{x}^*) \leq D(\mathbf{x}^*)$$

As each trade increases total profit, no feasible allocation that fails to satisfy (3-20) can maximize profit. Problem 12 sketches a simple proof that any feasible allocation $\mathbf{x}^*$ that satisfies (3-20) does maximize profit, so (3-20) is a necessary and sufficient condition for a feasible allocation to maximize profit.

Exercise 13. Check that the data in Table 3-2 satisfies (3-19). Using this data, initialize marginal analysis with the allocation $\mathbf{x} = (0, 4, 3, 2)$ and execute the trading step until you find the optimal solution $\mathbf{x}^* = (0, 3, 3, 3)$.

At each iteration of the trading step, one must compute $I(\mathbf{x})$, which is the largest forward difference. This can be done with N comparisons. However, at

Table 3-2. The data $p_n(x)$ for $n = 1, \ldots, 3$ and $x = 0, \ldots, 9$

					x					
n	0	1	2	3	4	5	6	7	8	9
1	0	6	12	18	20	22	24	25	20	10
2	0	4	8	12	15	18	18	18	18	18
3	0	6	12	18	20	15	10	5	0	-5

most two forward differences change from each iteration to the next, and the set of forward differences can be maintained as a *heap* (see Supplement 1) with the largest on top. The effort required to maintain the heap is proportional to $\log_2 N$. Similarly, the set of forward differences of which $D(\mathbf{x})$ is the smallest can be stored as a heap with the smallest on top.

Problem 13 demonstrates that marginal analysis has the following interesting property. If a variable is increased (alternatively, decreased) at any given iteration, it will not be decreased (alternatively, increased) at any subsequent iteration. So the variable x_n that is increased at a given iteration can be removed for all subsequent iterations from the heap having $D(\mathbf{x})$ on top! Similarly, the variable x_m that is decreased at a given iteration can be removed from the heap having $I(\mathbf{x})$ on top.

Another consequence of this property becomes evident when one considers the way in which marginal analysis moves from a feasible allocation $\mathbf{x}$ to an (optimal) allocation $\mathbf{x}^*$ that satisfies (3-20). At each nonterminal iteration of the trading step, exactly one x_n that exceeds x_n^* is decreased, and exactly one x_m that lies below x_m^* is increased. Hence, with $(z)^+ = \max\{0, z\}$, the number of nonterminal iterations of the trading step is precisely

$$\sum_{n=0}^{N} (x_n - x_n^*)^+$$

This sum is at most K. (Why?) So marginal analysis terminates after at most $K + 1$ executions of the trading step!

To compare marginal analysis with recursive fixing, we suppose that marginal analysis is initiated with the feasible allocation $(K, 0, \ldots, 0)$. Only (the phony) commodity 0 has production level above the optimum. Other production levels are increased only if doing so increases profit. Marginal analysis decreases x_0 at each iteration and increases by one the production of the commodity whose marginal contribution to profit is greatest. The quantity $D(\mathbf{x})$ is initially zero and remains zero. There is, then, no need to compute $D(\mathbf{x})$ at any iteration. Computing $I(\mathbf{x})$ in the most naive way entails comparison of N numbers. With at most $K + 1$ executions of the trading step, the number of quantities compared in this implementation of marginal analysis is at most $N(K + 1)$. The number of quantities compared in recursive fixing is exactly $N(K + 1)(K + 2)/2$, which is greater by the *factor* $(K + 2)/2$.

Exercise 14. Solve Program 4 for all values of K between 0 and 9, using the data in Table 3-2, and starting with the allocation $\mathbf{x} = (9, 0, 0, 0)$ (e.g., $x_0 = 9$).

Not only is marginal analysis faster; it can be initiated with any feasible allocation. In particular, it can be initiated with the decision maker's guess as to a good allocation, if he or she has one. Recursive fixing cannot make effective use of that kind of information.

Marginal analysis works so much better than recursive fixing that one is tempted to try it even when one or more of the profit functions are not quite concave. Suppose, for instance, that p_1 has the nonconcave shape displayed in Figure 3-2. Figure 3-2 also displays the *least concave majorant* q_1 of p_1, this being the lower envelope of the concave functions q that satisfy $q(x) \geq p_1(x)$ for all $x \geq 0$.

FIGURE 3-2. Least concave majorant q_1 of p_1

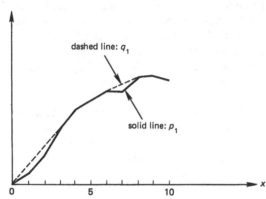

Imagine that we implement marginal analysis with p_1 replaced by q_1. Let $\mathbf{x}^*$ and $\mathbf{y}$ denote, respectively, the optimal allocations for the original allocation problem and for the contrived one. Problem 15 shows, in particular, that

$$0 \leq \sum_n [p_n(x_n^*) - p_n(y_n)] \leq q_1(y_1) - p_1(y_1)$$

So, if y_1 is a point (such as 5) at which the two functions coincide, then the solution of the contrived problem must be the solution of the original. If not, the preceding bound may still suffice. In many applications, exact solutions are not demanded, and one is willing to accept approximations like this, especially when the alternative is a great deal more computation and expense. Problems 16 and 17 are instances in which marginal analysis leads to an exact solution, and Problem 19 is an instance in which it provides a useful, approximate solution.

The discussion of marginal analysis is closed with a brief look at the "continuous" variant of Program 4, for which each x_n can be any nonnegative number, not just an integer. The continuous analogue of (3-19) is that each profit function p_n be concave on the set of nonnegative numbers and continuous at 0. This suffices (see Lemma 2 of Supplement 2) for existence of left and right derivatives $p_n^-(x)$ and $p_n^+(x)$, with

$$p_n^-(x) \geq p_n^+(x), \qquad \text{all } x > 0$$

The first forward difference is the discrete analogue of the derivative, and so $I(x)$ plays the role of the largest right derivative, while $D(x)$ plays the role of the smallest left derivative. This means that the analogue of (20) for the continuous case is that

(3-21)
$$\max_n \{p_n^+(x_n)\} \leq \min_{n \mid x_n > 0} \{p_n^-(x_n)\}$$

Problem 18 demonstrates that a feasible allocation x is optimal if and only if it satisfies (3-21).

ONE LAGRANGE MULTIPLIER*

Program 4 can also be attacked with a different weapon. The only constraint that links its variables is

(3-22)
$$x_0 + x_1 + \ldots + x_N = K$$

The left-hand side of this constraint can be multiplied by the scalar $-\lambda$, where λ is called a *Lagrange multiplier*, and shifted to the objective function. One then maximizes the expression

$$p_0(x_0) + \ldots + p_N(x_N) - \lambda[x_0 + \ldots + x_N]$$

subject (solely) to the constraint that each x_n be a nonnegative integer. For a *fixed* scalar λ, this separates into $N + 1$ subproblems, which are to maximize $\{p_n(x_n) - \lambda x_n\}$ for each n. These subproblems are easily solved, but the following questions arise. Does there exist a λ for which the solutions to the subproblems constitute a solution to Program 4? If so, how can we find this value of λ?

Our examination of these questions will reveal the essence of the theory of Lagrange multipliers. However, the specific detail of Program 4 is not relevant to this, and we shift our attention to a slightly more general context. Consider the allocation model having two resources, as defined by Program 3. The program stated below results from shifting the constraint on the second resource to the objective.

Program 5: Maximize $\{p_1(x_1) + \ldots + p_N(x_N) - \lambda[d_1(x_1) + \ldots + d_N(x_N)]\}$, subject to the constraints

(3-23)
$$c_1(x_1) + \ldots + c_N(x_N) \leq K$$

(3-24)
$$x_n \in \{0, 1, \ldots, B\} \qquad \text{for } n = 1, \ldots, N$$

---

* This section concerns a special topic that can be omitted without loss of continuity.

Note that L, which is the number of units of the second type of resource on hand, goes unmentioned in Program 5. Note also that Program 5 differs from Program 1 only in that the reward for producing x units of commodity n is changed from $p_n(x)$ to $p_n(x) - \lambda d_n(x)$. This makes it clear that for a *fixed* λ, Program 5 can be solved by reaching or recursive fixing.

Exercise 15. Write a functional equation appropriate to Program 5.

Let $\mathbf{x} = (x_1, \ldots, x_N)$ be any N-tuple of real numbers. Let $\mathbf{X}$ be the set of N-tuples $\mathbf{x}$ that satisfy (3-23) and (3-24), and note that $\mathbf{X}$ contains only finitely many members. So, for each fixed λ, the objective of Program 5 is maximized by some N-tuple $\mathbf{x}^\lambda$ in $\mathbf{X}$ that, as the notation indicates, can vary with λ. For each $\mathbf{x}$ in $\mathbf{X}$, let

$$p(\mathbf{x}) = p_1(x_1) + \ldots + p_N(x_N)$$
$$d(\mathbf{x}) = d_1(x_1) + \ldots + d_N(x_N)$$

LEMMA 3

(3-25) $\qquad p(\mathbf{x}^\lambda) = \max \{p(\mathbf{x}) | \mathbf{x} \in \mathbf{X}, d(\mathbf{x}) \leq d(\mathbf{x}^\lambda)\}, \qquad$ all $\lambda \geq 0$

(3-26) $\qquad d(\mathbf{x}^\alpha) \geq d(\mathbf{x}^\beta), \qquad\qquad\qquad\qquad\qquad$ all $\alpha < \beta$

Remark: Expression (3-25) indicates that $p(x^\lambda)$ is the largest profit attainable while consuming $d(\mathbf{x}^\lambda)$ or fewer units of the second resource. Consequently, if we are lucky enough to guess a value of λ satisfying

(3-27) $\qquad\qquad\qquad\qquad\qquad d(\mathbf{x}^\lambda) = L$

then a solution to Program 3 has been found. Expression (3-26) indicates that $d(\mathbf{x}^\lambda)$ is a nonincreasing function of λ, which hints at schemes for "homing in" on a value of λ satisfying (3-27).

Proof: Since $\mathbf{x}^\lambda$ is optimal for Program 5,

$$p(\mathbf{x}^\lambda) - \lambda d(\mathbf{x}^\lambda) \geq p(\mathbf{x}) - \lambda d(\mathbf{x}), \qquad \text{all } \mathbf{x} \in \mathbf{X}$$

Rearrange the above as

(3-28) $\qquad\qquad p(\mathbf{x}^\lambda) - p(\mathbf{x}) \geq \lambda[d(\mathbf{x}^\lambda) - d(\mathbf{x})], \qquad \text{all } \mathbf{x} \in \mathbf{X}$

Suppose that $\lambda \geq 0$. To verify (3-25), note that any $\mathbf{x} \in \mathbf{X}$ having $d(\mathbf{x}) \leq d(\mathbf{x}^\lambda)$ renders the right-hand side of (3-28) nonnegative; consequently, $p(\mathbf{x}^\lambda) \geq p(\mathbf{x})$.

To verify (3-26), first apply (3-28) with $\lambda = \alpha$ and $\mathbf{x} = \mathbf{x}^\beta$.

$$p(\mathbf{x}^\alpha) - p(\mathbf{x}^\beta) \geq \alpha[d(\mathbf{x}^\alpha) - d(\mathbf{x}^\beta)]$$

Now apply (3-28) with $\lambda = \beta$ and $\mathbf{x} = \mathbf{x}^\alpha$.

$$p(\mathbf{x}^\beta) - p(\mathbf{x}^\alpha) \geq \beta[d(\mathbf{x}^\beta) - d(\mathbf{x}^\alpha)]$$

Add the preceding two inequalities together.

$$0 \geq (\alpha - \beta)[d(\mathbf{x}^\alpha) - d(\mathbf{x}^\beta)]$$

So, when $\alpha < \beta$, it must be that $d(\mathbf{x}^\alpha) - d(\mathbf{x}^\beta) \geq 0$, which verifies (3-26). ∎

Examine Program 5 with $\lambda = 0$ to see that $p(\mathbf{x}^0)$ is the largest profit obtainable from any $\mathbf{x}$ in $\mathbf{X}$. If it happens that $d(\mathbf{x}^0) \leq L$, then the solution to Program 3 has been found, and the constraint $d(\mathbf{x}) \leq L$ is superfluous.

Bisection

Throughout this subsection, assume that $d(\mathbf{x}^0) > L$. Lemma 3 shows that any λ that satisfies (3-27) solves Program 3. Bisection is a simple method for finding such a λ. Start with a value c of λ large enough that $d(\mathbf{x}^c) < L$ and with a value b of λ small enough that $d(\mathbf{x}^b) > L$. Since $d(\mathbf{x}^\alpha)$ is a nonincreasing function of α, any value λ that satisfies (3-27) must lie between b and c. Try the midpoint; that is, solve Program 5 with $\lambda = (b + c)/2$. If $d(\mathbf{x}^\lambda) = L$, stop; a solution to Program 3 has been found. If $d(\mathbf{x}^\lambda) < L$, replace c by λ; if $d(\mathbf{x}^\lambda) > L$, replace b by λ. This *bisects* the interval in which any solution to (3-27) must lie. One can examine the midpoint again, iterating as follows.

Procedure:

1. Solve Program 5 with values b and c of λ designed to assure $d(\mathbf{x}^b) > L$ and $d(\mathbf{x}^c) < L$.

2. Solve Program 5 for $\lambda = (b + c)/2$. Stop if $d(\mathbf{x}^\lambda) = L$.

3. If $d(\mathbf{x}^\lambda) < L$, then $c \leftarrow \lambda$; otherwise, $b \leftarrow \lambda$. Go to step 2.

When does bisection work? This is best illustrated diagrammatically. In Figure 3-3, each element $\mathbf{x}$ of $\mathbf{X}$ is represented as a point in the plane, with coordinates $d(\mathbf{x})$ and $p(\mathbf{x})$. The five points above the "2" on the abscissa depict the five elements $\mathbf{x}$ having $d(\mathbf{x}) = 2$. Each of these five points has ordinate $p(\mathbf{x})$, and the topmost gives the largest profit obtainable while consuming exactly

FIGURE 3-3. The elements of $\mathbf{X}$—a well-behaved case

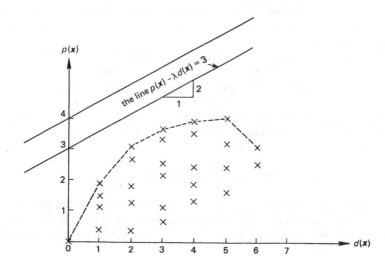

two units of the second resource. A dashed line connects the topmost points, and it is concave. As a function of L, the profit exhibits strictly decreasing marginal return. In this sense, the curve in Figure 3-3 is "well behaved."

The object of Program 5 is to maximize $p(\mathbf{x}) - \lambda d(\mathbf{x})$. In terms of Figure 3-3, this quantity is unchanging along straight lines with slope λ. When $\lambda = 0$, $p(\mathbf{x})$ is maximized without regard to $d(\mathbf{x})$, and the solution $\mathbf{x}^0$ to Program 5 has $d(\mathbf{x}^0) = 5$. As λ increases from 0, the lines along which it is constant tilt counterclockwise, and $d(\mathbf{x}^\lambda)$ takes, successively, the values 5, 4, 3, 2, 1 and then 0, each for an interval. Figure 3-4 illustrates this observation.

FIGURE 3-4. $d(\mathbf{x}^\lambda)$ vs. λ, a well-behaved case

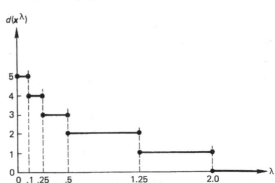

So, if L exceeds 4, $\mathbf{x}^0$ is optimal and the second constraint in Program 3 is superfluous. If $L \leq 4$, bisection eventually discovers a λ in the interval for which $d(\mathbf{x}^\lambda) = L$, which solves Program 3. Moreover, this should occur quickly, as the interval (b, c) is halved at each iteration. Actually, bisection is "minimax" in the sense that it obtains at each iteration the greatest possible guaranteed reduction of $c - b$.

In general, bisection works when the second resource contributes *strictly* decreasing marginal return. In that case, the dashed line bends downward at each integer, and a diagram like Figure 3-4 arises. To see what can go wrong, consider the ill-behaved case in Figure 3-5. As λ is increased from 0, the lines along which $p(\mathbf{x}) - \lambda d(\mathbf{x})$ is constant tilt counterclockwise, and $d(\mathbf{x}^\lambda)$ takes the values 7, 5, 3, and 0 in succession. The function $d(\mathbf{x}^\lambda)$ avoids the values 6, 4, 2, and 1. These points are called *gaps*.

Exercise 16.   Adapt Figure 3-4 to the ill-behaved case.

The gaps are now investigated. Consider Program 3 for the case $L = 6$. No allocation $\mathbf{x}$ has $d(\mathbf{x}) = 6$, and the solution to Program 3 is the topmost point $\mathbf{x}$ having $d(\mathbf{x}) = 5$. Bisection finds values b and c of λ for which $d(\mathbf{x}^b) = 7$

FIGURE 3-5. The elements of X—an ill-behaved case

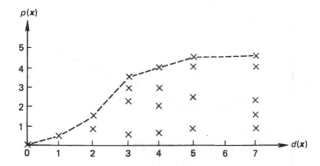

and $d(\mathbf{x}^c) = 5$. Then it continues to split the difference, without ever discerning that $\mathbf{x}^c$ solves Program 3. So the optimal solution is found, but not recognized.

Now suppose that $L = 4$. The optimal solution to Program 3 is the topmost point having $d(\mathbf{x}) = 4$, which solves Program 5 for exactly one value of λ and, even then, is not the unique solution. Barring extremely good luck, bisection finds values b and c of λ for which $d(\mathbf{x}^b) = 5$ and $d(\mathbf{x}^c) = 3$ and then iterates indefinitely, without ever finding the optimum.

Finally, suppose that L equals 1 or 2. The slope of the dashed line is increasing at these values, and no value of λ has $d(\mathbf{x}^\lambda)$ equal to 1 or 2. Bisection finds values b and c having $d(\mathbf{x}^b) = 0$ and $d(\mathbf{x}^c) = 3$, and then iterates indefinitely.

Updating the Slope

Should there be a gap at L, bisection iterates indefinitely. To set the stage for a method that always terminates finitely, suppose that Program 5 has been solved for values b and c of λ that satisfy

$$d(\mathbf{x}^b) > L > d(\mathbf{x}^c)$$

Examine Figure 3-6, in which λ is taken as the slope of the line connecting the points for b and c. The objective $p(\mathbf{x}) - \lambda d(\mathbf{x})$ of Program 5 is constant along this line, and its new solution $\mathbf{x}^\lambda$ necessarily has

(3-29) $$p(\mathbf{x}^\lambda) - \lambda d(\mathbf{x}^\lambda) \geq p(\mathbf{x}^c) - \lambda d(\mathbf{x}^c)$$

Since $\mathbf{x}^b$ and $\mathbf{x}^c$ maximize the objective of Program 5 when the Lagrange multiplier equals b and c, respectively, each element $\mathbf{x}$ of X must have point $[d(\mathbf{x}), p(\mathbf{x})]$ that lies on or below the dashed lines in Figure 3-6 that have slopes b and c. Similarly, since $\mathbf{x}^b$ and $\mathbf{x}^c$ are in X, the element $\mathbf{x}^\lambda$ must have $[d(\mathbf{x}^\lambda), p(\mathbf{x}^\lambda)]$ on or above the line in Figure 3-6 that has slope λ. Hence, $[d(\mathbf{x}^\lambda), p(\mathbf{x}^\lambda)]$ lies in the shaded region of Figure 3-6. Also, if (3-29) holds as an equality, there must be a gap at L. If (3-29) holds as strict inequality, $d(\mathbf{x}^\lambda)$ must lie *strictly* between $d(\mathbf{x}^c)$ and $d(\mathbf{x}^b)$. The process of updating the slope can be iterated, as follows.

Figure 3-6. Interpolation

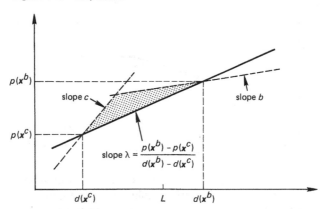

Procedure:

1. Solve Program 5 with values b and c of λ designed to assure that $d(\mathbf{x}^b) > L$ and $d(\mathbf{x}^c) < L$.

2. Solve Program 5 with $\lambda = [p(\mathbf{x}^b) - p(\mathbf{x}^c)]/[d(\mathbf{x}^b) - d(\mathbf{x}^c)]$. Stop if $p(\mathbf{x}^\lambda) - \lambda d(\mathbf{x}^\lambda) = p(\mathbf{x}^c) - \lambda d(\mathbf{x}^c)$.

3. If $d(\mathbf{x}^\lambda) < L$, then $c \leftarrow \lambda$; otherwise, $b \leftarrow \lambda$. Go to step 2.

Each nonterminal iteration of this procedure increases $d(\mathbf{x}^c)$ or reduces $d(\mathbf{x}^b)$, which assures finite termination. If the procedure stops with $d(\mathbf{x}^\lambda) = L$, then $\mathbf{x}^\lambda$ is optimal for Program 3. If the procedure stops with $d(x) \neq L$, then (1) there is a gap at L, (2) $\mathbf{x}^c$ is feasible for Program 3, and (3) the optimal solution $\mathbf{x}^*$ for Program 3 must satisfy

$$p(\mathbf{x}^*) \leq p(\mathbf{x}^c) + c[L - d(\mathbf{x}^c)]$$

the above being a direct application of (3-28).

At each iteration, bisection guarantees to halve the interval (b, c) in which λ might lie. Updating the slope guarantees finite termination. The following procedure combines the advantages of both methods. If $d(\mathbf{x}^b) - d(\mathbf{x}^c)$ is reduced on any given iteration, bisect on the next. If not, update the slope.

When there are no gaps, λ can be interpreted as a price. Observe in Figure 3-4 that $d(\mathbf{x}^\lambda) = L$ for all λ in an interval $[\lambda^-, \lambda^+]$. Notice in Figure 3-3 that increasing L by 1 unit increases profit by λ^-; that is, λ^- is the marginal value of the next unit of the second type of resource. Similarly, λ^+ is the marginal value of the Lth unit of the second type of resource. So λ^- and λ^+ can be interpreted as the decision maker's break-even buying and selling price for 1 unit of the second type of commodity. This sort of sensitivity information is often of interest in applications.

Exercise 17. Adapt the method of updating the slope to compute λ^- and λ^+.

LINEAR PROGRAMMING AND LAGRANGE MULTIPLIERS*

The preceding is an atypical introduction to Lagrange multipliers, which are usually described in the context of problems having continuous variables and several constraints. A typical development is to multiply each constraint by a "multiplier," shift all constraints to the objective, differentiate the objective with respect to all variables and multipliers, set all derivatives equal to zero, and then try to solve the resulting system of equations. That approach is sometimes successful, but it is very rarely insightful.

This section concerns multiple Lagrange multipliers. The ideas in Lemma 3 will be generalized, and they will form the key insight. Let us examine, again with $\mathbf{x} = (x_1, \ldots, x_N)$, the following optimization problem.

Program 6:   Maximize $p(\mathbf{x})$, subject to the constraints

$$a_i(\mathbf{x}) \leq b_i, \qquad i = 1, \ldots, m$$

$$\mathbf{x} \text{ in } \mathbf{X}$$

Introduce the m-tuple $\lambda = (\lambda_1, \ldots, \lambda_m) \geq 0$ of nonnegative Lagrange multipliers and consider, with λ fixed,

Program 7:   Compute $\phi(\lambda) = \max\limits_{\mathbf{x} \text{ in } \mathbf{X}} \{p(\mathbf{x}) - \sum\limits_{i=1}^{m} \lambda_i a_i(\mathbf{x})\}$.

In the preceding section, a specific structure was imposed on $\mathbf{X}$. No such structure is imposed here, but we do require that the optima to Programs 6 and 7 be attained. We shall attack Program 6 by repeated solution of Program 7 with various m-tuples λ of Lagrange multipliers. Of course, this approach is only of practical interest when $\mathbf{X}$ has enough structure that Problem 7 can be solved quickly (although not necessarily by dynamic programming).

Let us first fix $\lambda = (\lambda_1, \ldots, \lambda_m) \geq 0$ and examine the solution of Program 7. Let $\mathbf{x}^\lambda$ be an N-tuple in $\mathbf{X}$ that attains $\phi(\lambda)$. The following is analogous to Lemma 3.

LEMMA 4.   If $\lambda \geq 0$, then $p(\mathbf{x}^\lambda)$ maximizes $p(\mathbf{x})$ over all $\mathbf{x}$ that satisfy

$$(3\text{-}30) \qquad \begin{cases} a_i(\mathbf{x}) \leq a_i(\mathbf{x}^\lambda) & \text{if } \lambda_i > 0 \\ a_i(\mathbf{x}) \text{ unrestricted} & \text{if } \lambda_i = 0 \end{cases}$$

Moreover, if $\alpha_i = \beta_i$ for all $i \neq j$ and $\alpha_j < \beta_j$, then

$$(3\text{-}31) \qquad a_j(\mathbf{x}^\alpha) \geq a_j(\mathbf{x}^\beta)$$

---

* This section is advanced and concerns a special topic that can be omitted without loss of continuity. Knowledge of linear programming is necessary for a thorough understanding of the subject matter, and the preceding section is used here as well.

Remark: Expression (3-30) indicates that x^λ is optimal to Program 7, provided that

(3-32) $\begin{cases} a_i(x^\lambda) = b_i & \text{whenever } \lambda_i > 0 \\ a_i(x^\lambda) \le b_i & \text{whenever } \lambda_i = 0 \end{cases}$

Readers familiar with linear programming will recognize (3-32) as the *complementary slackness* conditions. Expression (3-31) indicates that increasing λ_j can decrease $a_j(x^\lambda)$, but cannot increase it.

Exercise 18. Prove Lemma 4.

[**Hint:** Follow the pattern of the proof of Lemma 3, starting with

$$p(x^\lambda) - \sum_{i=1}^{m} \lambda_i a_i(x^\lambda) \ge p(x) - \sum_{i=1}^{m} \lambda_i a_i(x) \quad \text{all } x \text{ in } X.]$$

Let us now explore the idea of updating the slope. Suppose that Program 7 has been solved with m-tuples $\lambda^1, \lambda^2, \ldots, \lambda^n$ of Lagrange multipliers, yielding (respectively) solutions $x^{\lambda^1}, x^{\lambda^2}, \ldots, x^{\lambda^n}$. At issue is how to use this information to determine λ^{n+1}. To cut down the superscripts, let

(3-33) $\begin{cases} a_{ij} = a_i(x^{\lambda^j}), & i = 1, \ldots, m, \quad j = 1, \ldots, n \\ c_j = p(x^{\lambda^j}), & j = 1, \ldots, n \end{cases}$

So c_j and the m-tuple $\mathbf{a}_j = (a_{1j}, \ldots, a_{mj})$ are the data obtained by solving Problem 7 with the jth m-tuple λ^j of Lagrange multipliers.

Perhaps a picture will help us see how to pick λ^{n+1}. Figure 3-7 depicts the very simplest case of one Lagrange multiplier ($m = 1$) with three data points (as $n = 3$). We know from the preceding section that the slope at b_1 is approximated by the slope of the (dashed) line connecting data points 2 and 3. Let us now figure out how to express this as the solution of a linear program. A line

Figure 3-7. Approximating the slope at b_1 (with $m = 1$ and $n = 3$)

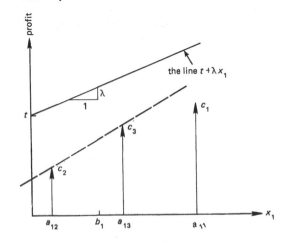

$t + \lambda x_1$ is defined in terms of its intercept t with the vertical axis and its slope λ with respect to the horizontal axis. The lines lying on or above the three data points are defined by the variables t and λ that satisfy

(3-34) $$t + \lambda a_{1j} \geq c_j, \qquad j = 1, 2, \ldots, 3$$

The height of a line above b_1 is $t + \lambda b_1$. Moreover, the dashed line in Figure 3-7 is the line that minimizes $t + \lambda b_1$ among those satisfying (3-34). This expresses the problem of approximating the slope as this linear program: minimize $\{t + \lambda b_1\}$, subject to (3-34). Its solution gives λ^{n+1}.

The case ($m = 2$) of two Lagrange multipliers is depicted in Figure 3-8,

FIGURE 3-8. Approximating the slope at **b** (with $m = 2$ and $n = 4$)

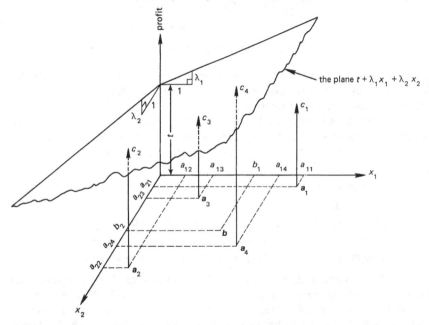

with four data points ($n = 4$). The slope above $\mathbf{b} = (b_1, b_2)$ is now approximated by a plane that intersects three of the data points and lies above the other. A plane is defined by its intercept t with the vertical axis and its slopes λ_1 and λ_2 with respect to the other two axes. That is, the expression

$$t + \sum_{i=1}^{2} \lambda_i x_i$$

defines a plane as a function of x_1 and x_2. Notice that planes lying on or above the four data points satisfy

(3-35) $$t + \sum_{i=1}^{2} \lambda_i a_{ij} \geq c_j, \qquad j = 1, \ldots, 4$$

The height of a plane above $\mathbf{b}$ is

(3-36) $$t + \sum_{i=1}^{2} \lambda_i b_i$$

The plane approximating the slope above $\mathbf{b}$ is the one that lies above the data points and whose height above $\mathbf{b}$ is minimal.

The general situation is to minimize an expression like (3-36) subject to constraints such as (3-35). In general, λ^{n+1} is specified as the vector of slopes that solve the following linear program.

Program 8: Compute $Q_n = \min \{t + \sum_{i=1}^{m} \lambda_i b_i\}$, subject to the constraints

(3-37) $$t + \sum_{i=1}^{m} \lambda_i a_{ij} \geq c_j, \quad j = 1, \ldots, n$$

$$\lambda_i \geq 0, \quad i = 1, \ldots, m$$

$$t \text{ unrestricted}$$

The constraint $\lambda_i \geq 0$ may come as a surprise. It reflects our prior knowledge, from (3-31), that $a_i(x^\lambda)$ is nonincreasing in λ_i. That assures us that only non-negative slopes need be considered.

In general, the procedure for updating the slope must be initialized with m different vectors $\lambda^1, \ldots, \lambda^m$ of Lagrange multipliers, selected so that $\mathbf{b}$ is a convex combination of $\mathbf{a}_1, \ldots, \mathbf{a}_m$. Thus initialized, the procedure iterates as follows.

Procedure for updating the slope:

1. Set $n = m$.
2. Compute λ^{n+1} as the optimal vector of slopes for Program 8.
3. Solve Program 7 with $\lambda = \lambda^{n+1}$. If x^λ satisfies (3-32), stop.
4. If $\phi(\lambda) = Q_n$, stop. Otherwise, $n \leftarrow (n + 1)$, and go to step 2.

If this procedure stops at step 3, the x^λ satisfying (3-32) is optimal for Program 6. If the procedure stops at step 4, no x^λ satisfies (3-32), and there is a gap at $\mathbf{b}$. In the case of a gap, we must settle for an approximate solution. Unfortunately, it can occur that x^λ and all of the active vectors in the final basis for Program 8 are infeasible for Program 6, although there is reason to hope that the extent of their infeasibility is small. When an infeasible solution is "rounded" to a feasible solution, Problem 20 bounds the loss of optimality.

Readers conversant with linear programming should recognize this as a *column generation* scheme. Program 7 generates the "column" $(\mathbf{a}_n, c_n)$ as the nth iteration. Program 8 turns out to require only one pivot per iteration. Moreover, the vector pivoted out is never pivoted back in and can be discarded.

Exercise 19. Interpret the constraints of Program 8 geometrically. Is $\mathbf{b}$ a convex combination of $\{\mathbf{a}_1, \ldots, \mathbf{a}_n\}$? What is the relation of $\mathbf{b}$ to the dual variables of Program 8?

Vinton, and Huntington, the last of which is currently used. Vinton's method was used from 1850 until the "Alabama Paradox" was observed.

Let $\lfloor x \rfloor$ denote the largest integer that does not exceed x (i.e., $1 = \lfloor 1.5 \rfloor = \lfloor 1 \rfloor$). Vinton's method for determining x_i is as follows. First allocate $\lfloor p_i H \rfloor$ seats to state i. Then allocate the remaining seats, one by one, in the following way. The state j having the largest remainder $p_i H - \lfloor p_j H \rfloor$ gets one, then the state for which that quantity is next largest gets one, and so on, until the stock of seats is exhausted. The following table shows the effect of this rule on a house of 100 seats. State 3 gets the extra seat.

State, i	$100p_i$	x_i	$101p_i$
1	45.3	45	45.75
2	44.2	44	44.64
3	10.5	11	10.61

a. What happens in the example described above when $H = 101$? This illustrates the "Alabama paradox," from which arose the principle that increasing the house size should not decrease the number of seats allocated to any state.

Huntington's method is based on this precept. Allocating $x_i + 1$ and x_j representatives to states i and j, respectively, is deemed preferable to allocating x_i and $x_j + 1$ representatives to these respective states if

$$\frac{p_i}{\sqrt{x_i(x_i + 1)}} > \frac{p_j}{\sqrt{x_j(x_j + 1)}}$$

b. Show that Huntington's method results in an allocation that minimizes the objective $\sum_i (p_i)^2/x_i$.

c. Show how to use marginal analysis to solve the problem described in part b. Is Huntington's method subject to the Alabama paradox?

d. A state i receiving fewer than $\lfloor p_i H \rfloor$ representatives may feel cheated. Can Huntington's method yield this outcome?

15. [*Approximate marginal analysis*] Consider Program 4, but without the hypothesis that each p_n is a concave function. Let $\mathbf{x}^* = (x_1^*, \ldots, x_N^*)$ be its optimal allocation. For each n, let q_n be a function (concave or otherwise) such that $q_n(x) \geq p_n(x)$ for every x. Let $y = (y_1, \ldots, y_N)$ be the optimal allocation for the variant of Program 4 in which, for each n, q_n replaces p_n. Show that

$$\sum_n p_n(y_n) \leq \sum_n p_n(x_n^*) \leq \sum_n q_n(x_n^*) \leq \sum_n q_n(y_n)$$

and, consequently, that

$$0 \leq \sum_n p_n(x_n^*) - \sum_n p_n(y_n) \leq \sum_n [q_n(y_n) - p_n(y_n)]$$

Exercise 20. Consider a program just like Program 3, but with six constraints. Suppose that constraints 1 to 3 are well behaved and constraints 4 to 6 are ill behaved. Which three would you put into the objective? Why?

BIBLIOGRAPHIC NOTES

Gilmore and Gomory (1966) may have been the first to accelerate reaching by exploiting special structure. They did this for the knapsack model of the "Reaching" section and for a multidimensional knapsack model akin to the allocation model in that section. Related computational experience is reported in Morin and Marsten (1976). The relation of knapsack models to integer programming is described in an excellent text by Garfinkel and Nemhauser (1972). For recent contributions to these subjects, see Denardo and Fox (1979a, 1979b, 1980).

The role of states in dynamic programming is systematically explored in Bellman (1957a), Denardo (1965), Elmaghraby (1973), and elsewhere.

Gross (1956) proposed marginal analysis for allocation models having decreasing marginal returns. His ideas were developed further by Fox (1966). Curiously, most of the "dynamic programming" examples that one finds in introductory operations research texts can be solved quicker by marginal analysis.

Lagrange multipliers have been an integral part of mathematics for nearly two centuries, but Everett (1963) is generally credited with suggesting their application to problems having integer variables. Brooks and Geoffrion (1966) discovered the relationship between Lagrange multipliers and linear programming. The section on linear programming and Lagrange multipliers draws heavily on these two sources. The section on one Lagrange multiplier uses Fox and Landi (1970). Some useful work has been done on resolving the "gaps," of which we cite Bellmore, Greenberg, and Jarvis (1970) and Shapiro (1979). For a recent review of Lagrangian relaxations of integer programs, see Fisher (1981).

PROBLEMS

1. Use recursive fixing to solve functional equation (3-6) with these data: $K = 3$, $t_{ij} = t_{ji}$, $t_{12} = 5$, $t_{13} = 4$, $t_{23} = 2$, $y_1 = 1$, $y_2 = 2$, $y_3 = 1$. Include with your solution to (3-6) an optimal poaching policy.

2. Write a functional equation for the following variant of Trout Rader's problem. He wishes to return to lake 1 as quickly as possible, bringing 10 trout, as his wife's recipe calls for poaching that many.

3. It is 10 hours until test time, and Dawn Grinder wishes to allocate them among the four subjects that constitute her dynamic programming course.

Allocating 0, 1, 2, 3, 4, and 5 hours to subject A will, she estimates, improve her score by 0, 2, 2, 3, 4, and 4 points, respectively. Allocating 0, 1, 2, 3, 4, and 5 hours to subject B will improve her score by 0, 1, 1, 2, 3, and 4 points, respectively. The yields for subjects C and D are, respectively, identical to those for subjects A and B. Write a functional equation whose solution yields her optimal study plan. Solve it by recursive fixing. What does she lose if she takes an hour off?

4. A corporation has K dollars available this year for investment in new projects. Its planning department contains L individuals. For $n = 1, \ldots, N$, new project n requires an investment of c_n dollars, and d_n planners are needed to undertake it. If project n is undertaken, the income it will earn in future years is equivalent to p_n dollars. Formulate the problem of computing the set of projects that should be undertaken for solution by dynamic programming.

5. Find an analogy between the concept of a state in dynamic programming and a *sufficient statistic* in statistics.

6. Show, in the knapsack model, how to find the smallest sack whose value is at least V.

7. Suppose, in the knapsack model, that there is an upper limit of B on the number of units of any single commodity that can be placed in the sack. Write an appropriate functional equation.

8. [Magazine, Nemhauser, and Trotter (1975)] Les Silver wonders how to make change using the fewest number of U.S. coins. He has numbered the coins in increasing value: $c_1 = 1$, $c_2 = 5$, $c_3 = 10$, $c_4 = 25$, and $c_5 = 50$.

   a. Write a functional equation whose solution $f(j)$ is the fewest number of U.S. coins whose total value is j cents.

   b. A *myopic* (or *greedy*) solution to this knapsack problem is, for each j, to pack the largest-valued coin n having $c_n \leq j$. Show that a myopic solution is optimal.

   [Hint: The inequalities $x_1 \leq 4$, $x_2 \leq 1$, $x_3 \leq 2$, $x_2 + x_3 \leq 2$, and $x_4 \leq 1$ suffice for $f(50 + j) = 1 + f(j)$, $f(25 + k) = 1 + f(k)$, and so on. Explain why.]

   c. Show (by example) that the myopic solution would not be optimal if the set of U.S. coins included a 20-cent piece.

   Note: Problems 9 to 11 relate to the section on reaching.

9. The Trout Rader example has enough structure to accord reaching an advantage over recursive fixing.

   a. Observe in (3-6) that $f(n, i) \geq f(n - 1, i)$. Show how to use this to accelerate reaching.

   [Hint: Check Lemma 2.]

   b. Apply accelerated reaching to the data in Problem 1.

10. [*Verification* of (3-8)] In equation (3-8), let $m(j) = n$ and $m(j - c_n) =$ Suppose that (3-8) is false or, equivalently, that $n > k$. So the final two a in the optimal path from node 0 to node j correspond to packing commod k, then commodity n. What is the effect of packing sequence on total valu Can n exceed k?

11. [*A bound on* i^*] In the knapsack problem, let $r = \max \{p_n/c_n | n \neq 1\}$. Suppo that p_n/c_n is maximized uniquely by $n = 1$, which means that $r < p_1/c$ Excluding item 1 from a sack having volume j results in a value of at mo jr. Excluding everything but item 1 from this sack produces value $p_1 \lfloor j/c_1$ $> p_1(j/c_1 - 1)$. Show that

$$i^* < \frac{p_1 c_1}{p_1 - c_1 r}$$

Note: Problems 12 to 18 relate to the discussion of marginal analysis in the section on decreasing marginal return.

12. [*That* $I(\mathbf{x}) \leq D(\mathbf{x})$ *suffices for optimality*] Let $s_n(a, b) = [p_n(b) - p_n(a)]/$ $(b - a)$. The function p_n is concave, and the proof of Lemma 2 of Supplement 2 shows that $s_n(a, b) \geq s_n(a, c) \geq s_n(b, c)$ whenever $a < b < c$.

   a. Let $\mathbf{x} = (x_1, \ldots, x_N)$ and $\mathbf{y} = (y_1, \ldots, y_N)$ be feasible for Program 4. Show that

   $$s_n(x_n, x_n + 1) \geq s_n(x_n, y_n) \qquad \text{if } x_n < y_n$$
   $$I(\mathbf{x})(y_n - x_n) \geq p_n(y_n) - p_n(x_n) \qquad \text{if } x_n < y_n$$
   $$s_n(x_n - 1, x_n) \leq s_n(y_n, x_n) \qquad \text{if } x_n > y_n$$
   $$D(\mathbf{x})(y_n - x_n) \geq p_n(y_n) - p_n(x_n) \qquad \text{if } x_n > y_n$$

   b. Now suppose that $I(\mathbf{x}) \leq D(\mathbf{x})$. Sum some of the above to obtain

   $$0 \geq \sum_n p_n(y_n) - \sum_n p_n(x_n)$$

13. [*The trading step*] Initialize marginal analysis with the feasible allocation $\mathbf{x}^0$ and let $\mathbf{x}^k$ be the allocation that results from the kth iterate of the trading step.

   a. Show that $I(\mathbf{x}^k) \leq I(\mathbf{x}^{k-1})$ and that $D(\mathbf{x}^k) \geq D(\mathbf{x}^{k-1})$.

   b. If variable x_n was decreased to $x_n - 1$ at some iteration and increased to x_n at some later iteration, it must be the case that $D(\mathbf{x}^r) = \Delta p_n(x_n - 1)$ and that $I(\mathbf{x}^s) = \Delta p_n(x_n - 1)$ for some $r < s$. Is this possible?

14. The House of Representatives has H seats (currently 435) to be allocated among N states (currently 50) "according to their respective numbers," in the words of the Constitution. Let p_i denote the fraction of the nation's population that resides in state i. The number of seats x_i allocated to state i must be an integer and, in particular, an integer close to $p_i H$. A recent book by Balinski and Young (1982) describes and analyzes congressional allocation methods proposed by Jefferson, Adams, Webster, Hamilton,

16. [*Approximate marginal analysis, applied*] Ward Herder has six helpers, who are anxious to assist him in turning out the vote. The following table contains his estimate of the increment in the vote that would result from assigning any number of helpers to each of the four precincts for which he is responsible.

Precinct	Number of Helpers						
	0	1	2	3	4	5	6
A	0	25	45	57	65	70	73
B	0	18	39	61	78	90	95
C	0	20	42	60	75	85	90
D	0	28	48	65	74	80	85

The problem is to assign helpers to precincts so as to maximize the sum of the increments in the vote. Do not solve this problem by dynamic programming. Instead, construct a table of first forward differences. Increase three numbers in the table in this text so that the forward differences become nondecreasing. (Figure 3-2 shows how to do that.) Then use marginal analysis on the altered data to do the allocation. Then use Problem 15 to see how much you might have missed the optimum by. Then do two further "trades" on the original forward difference to obtain the optimum.

17. E. Z. Rider has just arisen and has demoralized Dawn (Problem 3) by showing her that most of her calculations were unnecessary. What, exactly, did he show her?

18. Lemma 2 of Supplement 2 shows that if g is a concave function, then $s(a, b) \geq g^-(b) \geq g^+(b) \geq s(b, c)$ whenever $a < b < c$.

a. Suppose that x satisfies (3-21). Show that x is optimal.

[Hint: Mimic Problem 12.]

b. Suppose that x does not satisfy (3-21). Find a small trade that improves on x.

c. An optimum solution to Program 4 must exist (as a continuous function is being maximized over a closed bounded set). Show that (3-21) is necessary and sufficient for optimality.

19. Consider the optimization problem:

$$\text{Minimize} \left\{ \sum_{i=1}^{n} \sum_{j=1}^{m} B_{ij}(S_{ij}, S_{i0}) \right\}$$

subject to the constraints

$$\sum_{i=1}^{n} \sum_{j=0}^{m} c_i S_{ij} \leq K$$

$$S_{ij} \in \{0, 1, 2, \ldots\} \qquad \text{for } 1 \leq i \leq n, \quad 0 \leq j \leq m$$

This problem arises from a study conducted by The Rand Corporation for the U.S. Air Force. The Air Force maintains repair capabilities for roughly 2000 high-value items ($n \simeq 2000$), each of which is used at a number of bases ($m \simeq 20$). Should a type i item break at base j, it is repaired there, if possible. If the damage is too extensive for repair at the base, the item is sent to the depot responsible for it, where it is repaired or replaced. Stocks of spares are maintained at the depot and at the bases. Let S_{i0} denote the authorized stock level of item i at the depot, and let S_{ij} denote the authorized stock level of item i at base $j > 0$. The unit cost of item i is c_i dollars ($c_i > 0$), and the total budget for spare parts is roughly \$3 billion ($K \simeq 3 \times 10^9$). A "backorder" occurs if an item breaks for which no spare is immediately available. One reasonable criterion for allocating spares to the bases and the depot is to do so to minimize the average number of backorders in the system. In a simplified, but reasonable queueing model analyzed by Sherbrooke (1968), the average number of backorders occurring for item i at base j is a function $B_{ij}(S_{ij}, S_{i0})$ of the base stock S_{ij} and the depot stock S_{i0} of item i. Hence, the problem displayed above allocates stock to the bases and the depot so as to minimize the average number of backorders in the system.

This (backorder) function $B_{ij}(x, y)$ is (as one might suspect) decreasing in x and in y. It is also convex in x, and it is recursively computable in the sense that, for each y, the number $B_{ij}(x, y)$ can be computed successively for $x = 0, 1. \ldots$. The following approach to the allocation problem is a slick combination of marginal analysis and Lagrange multipliers that is due to Fox and Landi (1970). With $\lambda \geq 0$, consider:

(*)   Minimize $\left\{ \sum_{i=1}^{n} \left[\lambda c_i S_{i0} + \sum_{j=1}^{m} \left(\lambda c_i S_{ij} + B_{ij}(S_{ij}, S_{i0}) \right) \right] \right\}$

subject to $S_{ij} \geq 0$, integer for all i, j.

a. Show how to compute, efficiently,

$$g_{ij}(\lambda, x) = \min_{S_{ij}} \{ \lambda c_i S_{ij} + B_{ij}(S_{ij}, x) \}$$

$$g_i(\lambda) = \min_{x} \{ \lambda c_i x + \sum_{j=1}^{m} g_{ij}(\lambda, x) \}$$

b. How does the S_{ij} that attains $g_{ij}(\lambda, x)$ vary with λ? Why?

c. Put a grid $\lambda_1 > \lambda_2 > \ldots > \lambda_M > 0$ on the Lagrange multiplier. Show how to solve (*) successively for λ_1, λ_2, and so on, ending with λ_M.

d. Show how to find an approximate solution to the original problem from the solutions to the subproblems.

e. Interpret the near-optimal value of the Lagrange multiplier.

20. [Everett (1963)] Suppose, in Program 7, that one is able to find an m-tuple $\mathbf{x}'$ in $\mathbf{X}$ that comes within ϵ of attaining $\phi(\lambda)$. Adapt (3-30) to this situation. Verify it.

4

STAGES, GRIDS, AND DISCRETIZING CONTROL PROBLEMS

Titles above in bold type concern advanced
or specialized subjects. These sections can
be omitted without loss of continuity.

HOW STAGES ARISE

Nearly all introductory accounts of dynamic programming begin with discussions of stages. States are, however, more fundamental than stages. All models have states. Some have stages. Whether or not a model has stages can depend on how it is formulated, and the formulations without stages often yield faster computation of optimal policies than the ones with stages. These are the reasons why we introduced states first.

Still, stages are very important. The states in a sequential decision process often group themselves into stages, with transition occurring from each stage to the next. When this occurs, it is convenient to label each stage with an integer and to think of the process as evolving from stage 1 to stage 2, then to stage 3, and so on. A great advantage of stages is that they are easy to recognize. Recognizing the stages of a sequential decision process simplifies the task of formulating it for solution by dynamic programming.

A model having stages is analyzed in this chapter, and an illustration is provided of one of the ways in which stages arise. This illustration is a control problem in continuous time. Rarely can one compute the exact solution to a problem whose control can vary continuously with time. To approximate the optimal control, break time into discrete intervals, and allow the control to be adjusted at the start of each interval. Each stage in the discretized model contains the possible states of the system at the beginning of a time interval, and each transition increments time by one interval. If the time intervals are short enough, the optimal control for the discretized problem approximates the optimal control for the continuous-time problem.

Models having stages also arise from situations in which it is efficient to plan on a periodic basis. For instance, sales forecasts might be coordinated with inventory and production once each month. Chapter 5 concerns a production control problem whose stages reflect the times at which inventory and production are reviewed.

A CHAIN OF EVENTS

A sequential decision process entails a certain *chain of events* that is described in general and then particularized to models that have stages. A *state s* is an element of a set S called the *state space*. Associated with each state s is a *decision set* $D(s)$. The decision maker observes the state s (in S) that the system occupies. He or she then selects a decision d from the set $D(s)$. Selecting decision d for state s earns reward $r(s, d)$, which can be negative, and causes transition to state $t(s, d)$. The decision maker observes the state $t(s, d)$ that the system now occupies, and the chain of events repeats itself.

This chain of events is broken by termination. Termination occurs as soon as the system occupies a state s whose decision set $D(s)$ is empty. In addition, a decision d associated with a particular state can cause decision making to terminate, which we model by setting $t(s, d)$ equal to the empty set. Theoretically, either model of termination would suffice. Practically, both prove useful.

The model is said to *have stages* if some integer N exists such that the state space S can be partitioned into $N + 1$ sets $S_1, S_2, \ldots, S_{N+1}$ that are disjoint and that have the following properties. Decision set $D(s)$ is empty if and only if s is contained in S_{N+1}. Furthermore, one has $t(s, d)$ in S_{n+1} for all s in S_n, all $n \leq N$, and all d in $D(s)$. Hence, transition occurs from S_1 to S_2, then to S_3, and so on, finally to S_{N+1}.

A FUNCTIONAL EQUATION

When a model has stages, it is convenient to abuse notation by appending to states and decisions subscripts that indicate their stages. From a technical viewpoint, these subscripts are redundant. Practically, they are convenient. State s_n is automatically an element of S_n, and decision d_n is automatically a decision for some state in stage n. Figure 4-1 uses this notation to depict transitions from stage to stage.

The sequential decision process is incompletely specified because we have not yet stated an optimality criterion. The decision maker is now presumed to maximize the *sum* of the rewards he or she earns, and no rewards are associated with stage $N + 1$. To obtain a functional equation for this sequential decision process, we imagine that transition has occurred to some state s_n in stage n. How that happened is unrelated to the problem of selecting decisions d_n through

FIGURE 4-1. Evolution of a model having stages

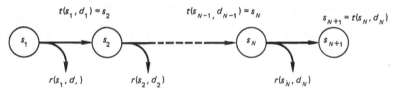

d_N to maximize the sum of the rewards earned at stages n through N. Consequently, we can define the optimal-return function f by

$f(s_n) =$ the maximum total reward obtainable at stages n through
$N + 1$ if state s_n is occupied

Selecting decision d_n for state s_n earns reward $r(s_n, d_n)$ and causes transition to state $t(s_n, d_n)$; also, $f[t(s_n, d_n)]$ is the best that can be obtained from that point on. This leads directly to the *functional equation*[1]

$$(4\text{-}1) \qquad f(s_n) = \begin{cases} 0, & n = N + 1 \\ \underset{d_n \text{ in } D(s_n)}{\text{maximum}} \{r(s_n, d_n) + f[t(s_n, d_n)]\}, & n < N + 1 \end{cases}$$

Recursive fixing consists of evaluating (4-1) for all states in stage N, then for all states in stage $N - 1$, and so on, ending with stage 1. This works because $t(s_n, d_n)$ is in stage $n + 1$.

Exercise 1. Suppose that transitions can occur from stage n to any higher-numbered stage. Does (4-1) still hold? Can recursive fixing be used?

Exercise 2. Suppose that the stage returns were multiplied together, not added, and suppose that $r(s, d)$ is nonnegative for all s and d. Does a principle of optimality still hold? Is there an analogue of (4-1)? Is $f(s_{N+1})$ still zero?

In Chapter 3, the section "Allocating One Type of Resource" describes an allocation model having one resource. To cast this model in a form that has stages, imagine that commodities 1 through N are produced in numerical order, commodity 1 first, and that commodity n is about to be produced. State (n, y) depicts the allocation of y units of resource to production of commodities n through N. Stage n consists of states (n, y) for $y = 0, 1, \ldots, K$. The decision set for state (n, y) consists of those production levels x of commodity n that do not use more than y units of resource; that is,

$$D(n, y) = \{x \mid c_n(x) \leq y\}$$

The reward $r[(n, y), x]$ for selecting decision x for state (n, y) is the profit for producing x units of commodity n; that is,

$$r[(n, y), x] = p_n(x)$$

[1] Equation (4-1) reflects the tacit assumption that suprema are attainable. A condition sufficient for this is that each decision set $D(s)$ contains finitely many elements.

If decision x is selected for state (n, y), there remain $y - c_n(x)$ units of the resource to allocate to production of commodities $n + 1$ through N; hence,

$$t[(n, y), x] = [n + 1, y - c_n(x)]$$

Since each transition increases n by 1, this model has stages. It is, however, incompletely specified.

Exercise 3. Complete the specification of the allocation problem as a model having stages by including provision for termination.

This allocation model illustrates an advantage and a disadvantage of stages. Recognizing that stage n depicts production of commodities n through N helps formulate this model for solution by dynamic programming. However, the alternative formulation of this model that is given in the section on reaching in Chapter 3 lacks stages and facilitates faster computation of longest paths.

INITIAL-VALUE AND FINAL-VALUE PROBLEMS

To reinterpret the model having stages as a longest-route problem in an acyclic network, take S as the set of nodes in this network. For each integer $n \leq N$, each state s_n and each decision d_n, create an arc (s_n, s_{n+1}) of length $r(s_n, d_n)$ where $s_{n+1} = t(s_n, d_n)$. Interpret $f(s_n)$ as the length of the longest of the paths from node s_n to any node in S_{N+1}.

Figure 4-2 depicts the network representation of a model having four stages. The quantity $f(s_2)$ is, for instance, the longest of the lengths of the paths from node s_2 to any of the three nodes in stage 4.

FIGURE 4-2. A network having four stages

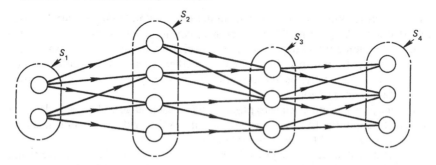

The network in Figure 4-2 has a property that is characteristic of models having stages; each path from any node in stage 1 to any node in stage n contains exactly $n - 1$ arcs, since arcs point from each stage to the next. We recall that the network analyzed in Chapter 2 had paths from node 1 to node 9 that contain three, four, and five arcs. Consequently, that network lacks stages.

Figure 4-2 makes it clear that we are discussing a family of *initial-value* problems; the path $(s_n, s_{n+1}, \ldots, s_{N+1})$ attaining $f(s_n)$ has node s_n fixed and node s_{N+1} unrestricted. Figure 4-2 is symmetric, and this suggests a family of *final-value* problems in which one computes the longest path to node s_n from any node s_1 in S_1. Let

$$F(s_n) = \text{the length of the longest of the paths}$$
$$\text{from nodes in } S_1 \text{ to node } s_n$$

These final-value problems can be solved by two familiar methods. To use recursive fixing, reverse all the arrows in Figure 4-2 and then apply the original analysis. To use reaching, compare

(4-2) $$F(s_n) + r(s_n, d_n)$$

with the best estimate found so far of $F[t(s_n, d_n)]$ and replace the latter by the larger. Repeat this step for every arc emanating from every node in stage n, first with $n = 1$, then with $n = 2$, and so on.

As computational methods, reaching and recursive fixing have comparative advantages and disadvantages. We have seen how to speed up reaching by exploiting the optimal policy's structure, but shall soon see that recursive fixing meshes better with interpolation. Notice that "reversing the arrows" amounts to finding for each s_{n+1} all pairs (s_n, d_n) such that $t(s_n, d_n) = s_{n+1}$. When the transition function is complex, reversing the arrows can entail troublesome functional inversions. Reaching avoids the need to reverse the arrows when solving final-value problems. Recursive fixing avoids the need to reverse the arrows when solving initial-value problems.

GRIDS AND GRID REFINEMENT

The natural formulation of a staged model often entails infinitely many states and decisions. Stage n might, for instance, consist of all pairs (n, y), where y ranges over some interval. Similarly, the decision variable d_n might vary continuously, say, from 0 to 1. In such cases, neither reaching nor recursive fixing can be applied directly, as these methods work only when there are finitely many states and decisions.

One possible approach is to place a *grid* on each stage and on each decision set and to apply the functional equation only to the grid points. For instance, y might be restricted to the values $0.00, 0.01, \ldots, 1.00$.

Let s_n and d_n be on the grids for stage n, and suppose that recursive fixing is used to compute $f(s_n)$. To evaluate the right-hand side of (4-1), we need to know the value of $f(s_{n+1})$, where

(4-3) $$s_{n+1} = t(s_n, d_n)$$

The difficulty is that when s_n and d_n lie on the grids for stage n, one can be all but certain that s_{n+1} lies off the grid for stage $n + 1$. When the grids are fine enough, $t(s_n, d_n)$ can be rounded to the nearest grid point. Otherwise, $f(s_{n+1})$ can be approximated by interpolating its values at the neighboring grid points. The latter technique is called *interpolation in the state space*.

Interpolating or rounding yields an approximation to $f(s_1)$. Now consider the problem of recovering a decision sequence $(d_1, \ldots, d_N)$ whose return approximates $f(s_1)$. While recursive fixing is being executed, it is a simple matter to record for each state its maximizing decision. Doing this produces d_1 routinely. The state s_2 in stage 2 to which transition occurs is $t(s_1, d_1)$. It is likely that s_2 lies off the grid for stage 2. One can round s_2 to the nearest grid point and take as d_2 its maximizing decision. Alternatively, one can interpolate the maximizing decisions for the neighboring grid points in the manner that was used to approximate $f(s_2)$. The latter of these approaches is called *interpolation in the policy space*. Whichever method is chosen must be applied repeatedly for decisions d_2 through d_N. The possibility of accumulated error is evident.

Now consider the effect of using grids with reaching. To avoid switching notation, we apply reaching to the final-value problems. The comparison associated with state s_n and decision d_n is to compute $s_{n+1} = t(s_n, d_n)$ and then replace the best estimate found so far of $F(s_{n+1})$ by the larger of itself and $F(s_n) + r(s_n, d_n)$. A difficulty arises from the fact that when s_n and d_n lie on the grids for stage n, state s_{n+1} is all but certain to lie off the grid for stage $n + 1$. One can still round s_{n+1} to the nearest grid point $\bar{s}_{n+1}$ and replace the best estimate found so far of $F(\bar{s}_{n+1})$ by the larger of itself and $F(s_n) + r(s_n, d_n)$. One cannot, however, interpolate because there is no natural place to store an improved estimate of $F(s_{n+1})$ when s_{n+1} is not a grid point.

If reaching is used, one can round to the nearest grid point. If recursive fixing is used, one can round or interpolate. That affords recursive fixing an advantage.

The staged model also illustrates the curse of dimensionality, together with a trick that sometimes overcomes it. Suppose that it is necessary to keep track of four variables in each stage. With a grid composed of 100 different levels of each variable, the number of grid points in stage n is $(100)^4$ or 10^6, which may be beyond the storage limit of the computer. A scheme that is aptly called *grid refinement* may help. Let us start with a rather "coarse" grid of, say, 10 levels of each variable and obtain a crude approximation to the optimal path from node s_1 to the end of the network. This reduces the number of nodes in each stage to 10^4, which is within the capability of today's computers. Then restrict attention to a neighborhood of this path, impose a finer grid on the neighborhood, and repeat the calculation.

One refines grids in the hope that the solution to the discretized problem approaches the solution to the original problem. This can be shown to occur when the reward and transition functions are continuous, jointly and uniformly, in the state and the decision. When grid refinement works, it provides an approximately optimal path from *one* node s_1 to any ending node, not from every beginning node. Also, the danger in starting with a coarse grid is that the optimal path may be outside the neighborhood of the path computed initially. When this happens, the optimal path cannot be approached by refinement of the grid.

SUMMARY

The preceding sections of this chapter complete an introduction to deterministic sequential decision processes. You should now be familiar with these ideas: states, transitions, rewards, embedding, the functional equation, recursive fixing, the principle of optimality, stages, grids, grid refinement, approximation in the state space, and approximation in the policy space. Of these ideas, states are the most central. The crucial property of states is this: Transition occurs from state to state; moreover, the transition function $t(s, d)$ and the reward function $r(s, d)$ can depend solely on the current state s and decision d.

Nearly everyone has difficulty at first in formulating optimization problems for solution by dynamic programming. Success at formulation comes, principally, from developing a knack for finding state spaces. An efficient way to find a state space is by trial and error. Start by guessing the stages and/or states. Check whether your guess leads to transition and reward functions that depend solely on the current state and decision. If so, you have succeeded. If not, revise the definition of the state variables, using the defects of your last attempt to guess more shrewdly. Then check again. After you succeed, write the functional equation. Then look closely at its boundary conditions (e.g., stage $N + 1$). Specifying the right boundary conditions can be tricky. It is often worth checking the boundary conditions by inventing a toy example and applying recursive fixing to it.

Assembled at the end of this chapter is a bouquet of formulation problems. They include some beauties. Your understanding of introductory dynamic programming will be complete when you can work most of these problems. Try. Remember to hunt for the state variables. Once you have found them, remember to check the boundary conditions.

The section that follows is starred. It concerns a control problem in continuous time. Classical approaches to this type of problem use the calculus of variations and often fall short of an exact solution. Our approach is to approximate the optimal control in a way that gives rise to a sequential decision process, then solve the sequential decision process by dynamic programming.

OPTIMAL CONTROL PROBLEM*

This section describes a method for approximating a control that can be adjusted continuously by a control that can be adjusted at discrete points in time. This method applies generally, but it is easier to understand when presented in

---

* This section concerns a special topic that can be omitted without loss of continuity.

a specific setting. We introduce it in the context of a chemical engineering problem.

Batch Reflux

A batch reflux is depicted in Figure 4-3. A fixed quantity of liquid, called the *batch*, is first placed in the primary vessel. A fire is then lit under the primary vessel. The fire raises the pressure in the primary vessel to the value for which the vessel was designed. The fire is adjusted, as necessary, to maintain this pressure. The liquid in the primary vessel boils and bubbles up a distillation tower. Its vapor passes through a valve, which splits it into two streams. A portion of the vapor is returned to the primary vessel for reheating, and the remainder is allowed to condense in the output pipe and drip into the output vessel. The ratio of returned vapor to output vapor is called the *reflux ratio*: hence the name *batch reflux*. The valve setting may be altered over time; it is the control variable.

FIGURE 4-3. A batch reflux in operation

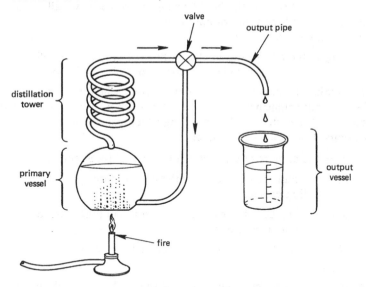

The batch reflux is a versatile device having many uses. We focus on its best-known application, the conversion of sweetened corn mash into liquor. In this context, the batch reflux is known as a *still*. We shall show how to operate the still to convert a batch of corn mash into any realizable quantity of liquor of any realizable alcoholic content. Our objective is to execute such conversions as quickly as possible. This objective ignores the quality of the liquor as well as the (scant) economic value of the portion of the batch that remains in the primary vessel at the end of processing.

Some Chemistry

The reactions that govern the evolution of this system are fermentation and differential boiling. Fermentation converts corn mash to alcohol. Different liquids boil at different rates, and this causes the alcoholic content of the vapor passing through the reflux valve to be higher than that of the liquid in the primary vessel.

The control variable in the batch reflux is the reflux ratio, which can and should vary over time. The technology governing the chemical evolution of the system is complex enough to preclude exact solution of the control problem in continuous time. So, as is often the case in control theory problems, one must settle for an approximation of the optimal control. The simplest sort of approximation arises from requiring the control to be piecewise constant. This approximates a continuous control, albeit crudely; it also approximates a discontinuous (*bang-bang*) control for which the standard methods of the calculus of variations are ill suited.

Dynamic Programming Formulation

It might seem natural to allow the reflux ratio to change at equally spaced intervals of time. However, the bookkeeping is simpler if we allow the reflux ratio to change each time a unit of output accumulates, and only then. Measure mass on a metric scale, and let

d_n = the reflux ratio in effect while the nth kilogram of output accumulates in the output pipe and vessel

Stage n contains the possible states of the system when accumulation of the nth kilo of output commences. The phrase "at epoch n" is now used to indicate the beginning of the interval of time during which collection of the nth kilo of output occurs. For each value of n, let

a_n = the number of kilos of alcohol in the still at epoch n

c_n = the number of kilos of unfermented corn mash in the still at epoch n

b_n = the number of kilos of alcohol in the output vessel and the output pipe at epoch n

M = the number of kilos of unfermented corn mash placed in the still initially

The number of kilos of fermented and unfermented corn mash in the still at epoch n is, by definition, $a_n + c_n$. At epoch n, exactly $n - 1$ kilos have flowed to the output pipe and vessel. Since mass is conserved,

$$(4\text{-}4) \qquad M = a_n + c_n + (n - 1)$$

Chemical engineering technology has determined three relevant functions, whose definitions and arguments are displayed below. All three functions depend on the composition (a_n, c_n) of the still at epoch n; two of them depend on the reflux ratio, d_n.

$t(a_n, c_n, d_n)$ = the time it takes to accumulate 1 kilo of output, given a_n, c_n, and d_n

$h(a_n, c_n)$ = the fraction (by weight) of the next kilo of output that is alcohol, given a_n and c_n

$g(a_n, c_n, d_n)$ = the number of kilos of corn mash fermented into alcohol during the time period between epochs n and $n + 1$

The amount of alcohol in the output pipe and vessel is increased by flow from the primary vessel. Hence,

(4-5) $$b_{n+1} = b_n + h(a_n, c_n)$$

The amount of alcohol in the still is increased by fermentation and decreased by outflow. Hence,

(4-6) $$a_{n+1} = a_n + g(a_n, c_n, d_n) - h(a_n, c_n)$$

The amount of corn mash in the still is decreased by fermentation and also by outflow. Hence,

(4-7) $$c_{n+1} = c_n - g(a_n, c_n, d_n) - [1 - h(a_n, c_n)]$$

A dynamic programming formulation is now easily identified. Clearly, d_n is the decision. A state must include the quantity b_n of alcohol in the output. Equation (4-4) interrelates the variables n, a_n, and c_n, only two of which need to be included as state variables. Let us take the triplet (n, a_n, b_n) as a state, noting that c_n is then determined by (4-4). For a network formulation, create one node for each triplet (n, a_n, b_n). For each valve setting d_n, create an arc emanating from node (n, a_n, b_n). The length of this arc is $t(a_n, c_n, d_n)$, and it terminates at node $(n + 1, a_{n+1}, b_{n+1})$, as specified in (4-5) and (4-6). Transition occurs from stage n to stage $n + 1$. The only node in stage 1 is $(1, 0, 0)$. Let

$F(n, a_n, b_n)$ = the length of the shortest path from node $(1, 0, 0)$ to node (n, a_n, b_n)

We have just constructed a family of final-value problems, where $F(n, a_n, b_n)$ is the shortest time in which the conditions specified by this triplet can be realized. Whatever remains in the still at the end of processing has no economic value and is discarded. Hence, we have no inherent interest in the final value of a_n. Define $G(n, b_n)$ by

$$G(n, b_n) = \min_{a_n} \{F(n, a_n, b_n)\}$$

Interpret $G(n, b_n)$ as the shortest time in which one can accumulate $n - 1$ kilos of output that include b_n kilos of alcohol.

Exercise 4. Since a_n is of no inherent interest, why not omit it from the definition of a state?

There was no compelling reason for selecting (n, a_n, b_n) as the state. Other triplets would do.

Exercise 5. Devise a formulation (i.e., describe the arcs) in which a triplet of the form (a_n, b_n, c_n) is a state.

With the formulation in Exercise 5, the decision variable d_n could be discretized at, say, 5-minute intervals. The resulting model would still have a direction of motion, as c_n is reduced continually by fermentation. So its network would be acyclic. Recursive fixing could still be used, but the model would not have stages. It would also accommodate leaving the valve fully closed, which might be optimal, especially at the beginning of distillation.

The variables a_n, b_n, c_n, and d_n all vary continuously, but each must be discretized to facilitate computation of shortest paths. Once the variables are discretized, the family of final-value problems can be solved by reaching or by recursive fixing. Recursive fixing accommodates a coarser grid because it meshes better with interpolation. However, reaching can be organized to evaluate arcs emanating only from nodes that can be reached from stage 1, and reaching can exploit whatever qualitative information is known about the optimal control.

One might combine the benefits of these methods by using recursive fixing for the initial approximation and reaching for later refinement. Start by placing coarse grids on the states, decisions, *and* the time axis; then use recursive fixing, with careful interpolation, to obtain an initial approximation to the optimal path. Next, restrict attention to a neighborhood of this path, refine all grids, and execute reaching, imposing whatever structure is known to exist on the optimal path.

Exercise 6. State (n, a_n, b_n) corresponds to $n - 1$ kilos of liquor whose alcoholic content, defined as fraction of total weight, is $b_n/(n - 1)$. Show how to find the policy that maximizes profit over a long planning horizon (e.g., 1 year) under these conditions: sweetened corn mash costs \$5 per batch, the still costs \$0.10 per unit time to operate, it takes 10 units of time to clean the still between batches, and each kilo of liquor of alcoholic content x is sold at price $p(x)$.

This batch reflux model illustrates many of the points discussed earlier. Grids are used to approximate continuous controls by discrete ones. Grids are seen to mesh better with recursive fixing than with reaching. Dynamic programming is seen to solve a family of optimization problems, not just one. In this instance, dynamic programming solves the family of final-value problems, each of which is of interest to the manager. The continuous-time control problem has been discretized in two different ways. Both ways yield shortest-path problems in acyclic networks. Only one of these networks has stages. The network that lacks stages is more flexible because it allows the reflux valve to be fully closed. Finally, dynamic programming methods approximate discontinuous, bang-bang controls that are hard to handle within the framework of the calculus of variations.

BIBLIOGRAPHIC NOTES

This chapter touches only lightly on the use of dynamic programming to study and solve deterministic control problems in continuous time. This topic is the subject of a survey by Bellman (1971) and a book by Larson (1968). Dreyfus (1966) investigates the relationship between dynamic programming and the calculus of variations. For an introductory account of this, see Dreyfus and Law (1978). The batch-reflux example is adapted from Mitten and Probhakar (1964).

PROBLEMS

1. Consider the optimization problem:

$$\text{Minimize } \{g(x_1) + \ldots + g(x_N)\}$$

subject to

$$x_1 + \ldots + x_N = L, \qquad x_i \geq 0, \qquad i = 1, \ldots, N$$

Suppose that $g(x)$ is differentiable and convex; that is, $g'(x)$ is nondecreasing in x. Show that $x_i \equiv L/N$ is optimal.

(Hint: Write a functional equation for $f(n, y)$, try induction, and differentiate the expression $g(x) + (n - 1)g[(y - x)/(n - 1)]$.)

2. Consider the optimization problem:

$$\text{Minimize } \{g(x_1) \cdot g(x_2) \ldots g(x_n)\}$$

subject to

$$x_1 + \ldots + x_N = L, \qquad x_i \geq 0, \qquad i = 1, \ldots, N$$

Suppose that $g(x)$ is positive and that $g'(x)/g(x)$ is increasing in x. Show that $x_i \equiv L/N$ is optimal.

[Hint: Take the logarithm and use Problem 1.]

3. Spaceship Vroom is being designed for a voyage to the nearest star, Alpha Centauri. A final payload of mass M_0 must be accelerated in free (zero gravity) space to $\frac{1}{10}$ the speed of light. The question is to design an n-stage rocket whose initial mass is minimal. Ninety percent of the mass of each stage is burned, and then the remainder is jettisoned before ignition of the next stage. Acceleration is governed by the Newtonian equation

$$M(t) \frac{d}{dt} V(t) = -\frac{d}{dt} M(t)$$

where $M(t)$ and $V(t)$ are the mass and velocity, respectively, of the spaceship at time t. (Coordinates are chosen so that the velocity of light is 1.)

Number the stages so that stage 1 is the last to burn. Let M_n denote the mass of the spaceship at the moment when stage n is ignited, and let x_n denote the increase in velocity due to stage n.

a. Integrate Newton's equation to obtain $x_n = \log(M_n) - \log(0.1M_n + 0.9M_{n-1})$.

b. Show that $M_n = M_{n-1} \cdot g(x_n)$, where $g(x) = 0.9e^x/(1 - 0.1e^x)$.

c. Let $f(n, v)$ be the smallest initial mass necessary to accelerate the final payload to velocity v using n stages. Write a functional equation for $f(n, v)$.

[Hint: Use part b.]

d. Show that $f(n, v)$ is attained by $x_i \equiv v/n$.

[Hint: See Problem 2.]

4. A long one-way street consists of m blocks of equal length. A bus runs "uptown" from one end of the street to the other. A fixed number n of bus stops are to be located so as to minimize the total distance walked by the population. Assume that each person taking an uptown bus trip walks to the nearest bus stop, gets on the bus, rides, gets off at the stop nearest his destination, and walks the rest of the way. During the day, exactly B_j people from block j start uptown bus trips, and C_j complete uptown bus trips at block j. (Of course, $\sum_{j=1}^m B_j = \sum_{j=1}^m C_j$.) Write a functional equation whose solution places the bus stops so as to minimize the total distance walked by the population of bus users. (Curiously, the data B_j and C_j suffice for this purpose; the number of people wishing to go from block i to block j need not be determined.)

5. Continuing the bus-stop problem, suppose that each block takes 3 minutes to walk. At full speed, the bus covers five blocks per minute. The bus takes 2 minutes to decelerate to a stop, unload passengers, load, and accelerate back to full speed. Assume that passengers walk to and from the bus stops nearest their points of origin and destination. Write a functional equation whose solution gives the number and location of bus stops that minimize the total travel time of the population.

6. Plants 1 through N are located along the Weladare River, with plant $n + 1$ downstream from plant n. A volume Q of pure water and $0.1Q$ of waste flows down the river, the waste coming from upstream sources. The water is to be consumed by the N plants. Each unit of water consumed by plant n produces a_n units of waste. The cost of consuming q units of water at plant n depends on the product of q and the quantity w of waste produced upstream; this is indicated by the cost function $c_n(q \cdot w)$. The problem is to allocate the water to the plants so as to minimize total regional cost. Set up a functional equation whose solution accomplishes this.

$$V_{n+1}(x, y) = \min\left\{ C_{n+1}(\text{?}, y) + V_n(x - q_{n+1}, y + a_{n+1}q_{n+1}) \right\}$$

$$V_0(x, y) = 0 \qquad \text{s.t.} \quad q_{n+1} \le x \qquad y \ge 0.1Q$$

$$x \le Q$$

7. A company has m jobs that are numbered 1 through m. Job i requires k_i employees, each of whose natural wage is w_i dollars per month, with $w_i \le w_{i+1}$. The jobs are to be grouped into n labor grades, a particular labor grade consisting of jobs $p + 1$ through q, inclusive. All employees in a given labor grade receive the highest of the natural wages of the jobs in that grade. A fraction r_j of the employees in each job quit in each month. Vacancies must be filled by promoting from the next lower grade. For instance, a vacancy in the highest of n labor grades causes $n - 1$ promotions and a hire into the lowest labor grade. It costs t dollars to train an employee to do any job. Write a functional equation whose solution determines the number n of labor grades and the set of jobs in each labor grade that minimizes the sum of the payroll and training costs.

8. A company must schedule regular-time and overtime production so as to meet demand for the next 10 months. The demand D_t for month t, the regular-time capacity R_t for month t, and the overtime capacity O_t are as given in the accompanying table. The unit production cost is higher using overtime capacity than regular-time capacity. An inventory carrying cost is incurred for each unit of inventory left over at the end of each month. These (unspecified) costs do not vary with the month. Write a recursive relation for $f(t)$, the best inventory to have at the beginning of month t. Solve it for the given data; do not try to justify or solve it for the general case.

	Month, t									
	1	2	3	4	5	6	7	8	9	10
Demand, D_t	13	11	13	12	10	8	11	11	11	10
Regular-time capacity, R_t	9	6	7	7	8	9	7	7	6	6
Overtime capacity, O_t	7	6	5	3	2	5	5	5	3	2

9. It is June 1, and No Wonder Bakers has 120 trained bakers in its employ. Company policy allows hiring bakers each month, but never allows firing them. Training each new baker takes 1 month and requires a trained baker to spend $\frac{1}{2}$ month acting as supervisor, rather than making bread for the company. The requirements for trained bakers who are productively employed are given below for the next 7 months.

Month	1	2	3	4	5	6	7
Number of trained bakers	100	105	130	110	140	120	100

$f(2) =$ 117, 132, 124 140 128 119
    12   3   26   0   0   0

$f(1) =$ 109 117   132   124 140 128 119

So the 20 extra trained bakers on hand in June can lay idle or train as many as 40 new bakers. Experience has shown that 8% of the trained bakers quit at the end of each month, but that none of the trainees does. The personnel director wants to minimize payroll costs. He wonders how many bakers to hire each month. Mercifully, No Wonder does not stock bread against future demand.

a. Write a functional equation involving $f(j)$, where $f(j)$ is the minimum number of trained bakers needed to satisfy the demand for months j through 7.

b. Solve the functional equation. How many employees should be hired each month?

10. One bomber and n tankers take off from one base, all fully loaded with fuel, all headed for a target, and all flying at the same speed. The fuel capacities of the bomber and tanker are 300 and 500 units, respectively. The bomber and tankers consume fuel at the rate of 0.1 unit per mile. When a tanker becomes empty, its crew bails out, and the plane is lost. Determine the maximum range of the bomber in these cases.

a. While aloft, the tankers can transfer fuel to each other and to the bomber.

b. While aloft, the tankers can transfer fuel to the bomber, but not to each other. Is there a significant disadvantage to part b, as compared to part a?

c. While aloft, the tankers can transfer fuel only to the bomber, and they must return to their home base.

11. Show how to use a Lagrange multiplier to reduce the dimension of the state space in the batch reflux problem.

12. The half-open line segment $[a, b)$ is to be defined by n soldiers against infiltration by one enemy agent. Each soldier is assigned a subsegment $[p, q)$ to defend. He positions himself at point r, where $p \le r < q$. The soldiers' subsegments do not overlap. If the agent attempts to penetrate at a point x in the segment defended by the soldier at point r, the agent succeeds with probability $P(x, r)$. Write a functional equation whose solution places the defenders so as to minimize the probability of infiltration, given that the agent knows where the defenders are.

    Remark: See Denardo, Huberman, and Rothblum (1979) for an "interleaving" property that Problems 4, 7, and 12 share.

13. A jet intercepter is to start at ground level and reach a velocity of 1800 mph at an altitude of 60,000 ft at quickly as possible. The relevant equation of motion is

$$\frac{dH}{dt} = \frac{\phi(H, V, \dot{H}, W)}{1 + (W/g)(dV/dh)}$$

where V is the jet's velocity in miles per hour, H its altitude in feet, W its weight, g the acceleration due to gravity, and ϕ is a known, slowly varying function of H, V, W, and $\dot{H} = dH/dt$.

a. Argue that the variables ΔH, ΔV, and Δt, when appropriately defined, are related to each other by

$$\Delta t = \frac{\Delta H + (W/g) \, \Delta V}{\phi(H, V, \Delta H/\Delta t, W)}$$

b. Place grids on H and on V at regular intervals of 1000 ft and 30 mph, respectively. Allow only two decisions: (1) increase velocity one grid point at constant altitude, and (2) increase altitude one grid point at constant velocity Write a functional equation whose solution gives the minimum time to climb.

> *Remark:* A slight variant of this example has been solved by Cartiano and Dreyfus, and the optimal trajectory [see Bellman and Dreyfus (1962, p. 216)] represents a control of the bang-bang sort, which would be difficult to obtain by classical methods.

c. The formulation in part b does not allow *dives*, which are increases in velocity at the expense of altitude, or *zooms*, which are increases in altitude at the expense of velocity. Thrust and lift at various speeds and altitudes are such that dives and zooms may be optimal. Still, it is reasonable to expect the optimal trajectory to increase the aircraft's energy monotonically, where the energy E is, as usual, defined by

$$E = \tfrac{1}{2}WV^2 + WH$$

Place a 60-point grid on E. Admit transitions that increase E by one grid point and that increase V by 2, 1, 0, -1, or -2 grid points. Write a functional equation whose solution is the minimal time to climb.

d. The prior discussion ignores fuel consumption, which decreases the weight W during climb. Let W_F denote the aircraft's weight, including fuel, at the point when it reaches 1800 mph at 60,000 ft. A significant portion of W_F is the fuel that will be consumed for operational purposes. Show how to find the minimum time to climb as a function of W_F, assuming that the rate of fuel consumption is a known, slowly varying function of H, V, $\dot{H}$, and W.

5

PRODUCTION CONTROL AND NETWORK FLOW

Titles above in bold type concern advanced
or specialized subjects. These sections can
be omitted without loss of continuity.

INTRODUCTION

This chapter begins with the study of the control of production over a multi-period planning horizon. The objective is to schedule production in the various periods so as to satisfy demand at minimum total cost. This planning problem is called *deterministic* because demand is assumed to be known in advance of planning. In the problems at this chapter's end, the model is reinterpreted in the contexts of capacity expansion and of inventory control.

Special attention is paid to the cases of economies of scale and of diseconomies of scale. In both cases, the cost-minimizing production schedule is seen to have interesting qualitative features.

Analysis of the production control problem leads naturally to the study of models of flows in networks. In the case of concave costs (economies of scale), one studies a network flow model whose special structure allows for efficient computation of an optimal flow. In the case of convex costs (diseconomies of scale), one studies a general model of network flow. This convex-cost network-flow model arises from our study of production control, but it has many other uses.

THE BASIC MODEL

Throughout, a *period* is an interval of time, and an *epoch* is the time at which a period begins. In particular, epoch n is the time at which period n starts.

The basic form of the production control model is now described. Demand for a single product occurs during each of N consecutive time periods that are numbered 1 through N. The demand that occurs during a given period can be satisfied by production during that period or during any earlier period, as inventory is carried forward. This prescribes the case of *no backlogging;* demand is not allowed to accumulate and be satisfied by future production. Inventory at epoch 1 (the start of planning) is zero, and inventory at the end of period N (the end of planning) is required to be zero. The model includes production costs and inventory carrying costs. The objective is to schedule production so as to satisfy demand at minimum total cost.

The data in this model are the demands, the production cost functions, and the inventory carrying cost functions. Exact specifications of these data are displayed below. For $n = 1, \ldots, N$,

$d_n = $ the demand during period n

$c_n(x) = $ the cost of producing x units of the product during period n

$h_n(y) = $ the cost of holding y units of inventory at epoch n [in the simplest case, the holding cost $h_n(y)$ is the cost of carrying for period n the investment in y units of inventory]

The decision variables are these: for $n = 1, \ldots, N$,

$x_n = $ the production during period n

$I_n = $ the inventory on hand at epoch n

Production and demand occur in integer quantities, and the problem of meeting demand at minimal total cost has the following mathematical representation.

Program 1.   Minimize $\sum_{n=1}^{N} \{c_n(x_n) + h_n(I_n)\}$, subject to the constraints

(5-1)	$I_1 = I_{N+1} = 0$		
(5-2)	$I_n + x_n = d_n + I_{n+1},$		$n = 1, \ldots, N$
(5-3)	$x_n \geq 0,$	x_n integer,	$n = 1, \ldots, N$
(5-4)	$I_n \geq 0,$	I_n integer,	$n = 2, \ldots, N$

In the above, (5-1) assures that initial and final inventories are zero. Matter is conserved, and (5-2) requires that the sum of the inventory at the start of a period and production during that period equal the sum of the demand during that period and the inventory at the start of the next period. Production and inventory are constrained by (5-3) and (5-4) to nonnegative integer values. The constraint $I_n \geq 0$ assures that demand in period n is satisfied by production during that period or during earlier periods.

The production and inventory levels are, of course, interrelated. If one knew the inventory levels at all the epochs, one could determine the production levels from (5-2). Conversely, if one knew the production levels x_1 through x_N, one could determine the inventory levels from the equations

(5-5) $\quad (x_1 + \ldots + x_{n-1}) = I_n + (d_1 + \ldots + d_{n-1}), \quad n = 2, \ldots, N$

To check (5-5), sum (5-2) from 1 through $n-1$ and cancel terms. To interpret (5-5), note that the inventory I_n at epoch n equals the total production during periods 1 through $n-1$ less the total demand during these periods. Call $(x_1, \ldots, x_N)$ a *feasible production plan* if it and the inventory levels determined by (5-5) satisfy the constraints of Program 1. Call $(x_1, \ldots, x_N)$ an *optimal production plan* if it is a feasible production plan that minimizes the objective of Program 1 over all feasible production plans.

A FUNCTIONAL EQUATION

To find an optimal production plan by dynamic programming, note that the cost of operating the system during periods n through N depends on the inventory I_n at epoch n, but not on prior inventories and not on prior production levels. So the pair (n, I_n) constitutes a state. Let

$f(n, I_n) =$ the minimum cost of satisfying demand during periods n through N if the inventory at epoch n is I_n.

The quantity $f(n, I_n)$ is intended to include the holding cost $h_n(I_n)$ that is incurred immediately as well as all future production and holding costs. The cost of an optimal production plan is $f(1, 0)$.

Final inventory must be zero, which constrains inventory and production during period N. If the inventory level I_N is below d_N, one must produce the difference. Inventory levels in excess of d_N are forbidden. Hence,

(5-6) $\quad f(N, I_N) = h_N(I_N) + c_N(d_N - I_N) \quad$ for $I_N = 0, 1, \ldots, d_N$

Consider a state (n, I_n) with $n < N$, and let x_n denote the quantity produced during period n. The sum $I_n + x_n$ of the inventory at epoch n and the production during period n must be at least as large as the demand d_n during that period. Also, the sum $I_n + x_n$ can be no larger than the total demand during all remaining periods. So production level x_n is called *feasible* for state (n, I_n) if x_n is a nonnegative integer satisfying

(5-7) $\qquad\qquad d_n \leq I_n + x_n \leq d_n + \ldots + d_N$

There are finitely many nonnegative integers x_n that satisfy (5-7). The usual argument gives rise to the functional equation

(5-8) $\quad f(n, I_n) = \min_{x \text{ feasible}} \{h_n(I_n) + c_n(x) + f(n+1, I_n + x - d_n)\}, \quad n < N$

The preceding functional equation can be solved by recursive fixing or by reaching.

If the inventory I_n and production x_n were allowed to vary continuously, (5-8) would still hold, but with "inf" replacing "min." In order to solve the continuous analogue of (5-8), one might need to impose grids on I_n and on x.

CONCAVE COSTS

Let J denote the set of nonnegative integers: $J = \{0, 1, 2, \ldots\}$. A real-valued function g is called *concave on J* if

(5-9) $g(x + 2) - g(x + 1) \le g(x + 1) - g(x)$,   all x in J

Interpret $g(x)$ as the cost of producing x items; (5-9) holds when the incremental cost producing item $x + 2$ is no greater than the incremental cost of producing item $x + 1$. This reflects a situation of *decreasing marginal costs*, or, equivalently, economies of scale. These terms are synonymous with concave costs. The model is said to have *concave production costs* if, for each n, the function c_n is concave on J. Similarly, the model has *concave holding costs* if, for each n, the function h_n is concave on J. When the production and holding costs are concave, the solution to Program 1 has the simple form given below.

THEOREM 1. (*concave costs*). Consider Program 1 for the case of concave production and holding costs. There exists at least one optimal production plan $(x_1, \ldots, x_N)$ for which

(5-10) $I_i x_i = 0$ for $i = 1, \ldots, N$

The proof of Theorem 1 is deferred to a later section, in which several results related to it are gathered. Equation (5-10) demonstrates that production need only occur in period i if the inventory at the start of that period is zero. This must also be true of the next period j at which production occurs. Consequently, the quantity produced during period i must equal the total demand during periods i through $j - 1$, which is defined below as d_{ij}.

$$d_{ij} = \sum_{k=i}^{j-1} d_k, \quad i < j \le N + 1$$

This gives rise to a dynamic programming formulation in which state i represents the situation of having no inventory on hand at the start of period i. Let

$f(i) =$ the minimal cost of satisfying demands during periods i through N if $I_i = 0$

Let c_{ij} denote the total of the costs incurred during periods i through $j - 1$ if transition occurs from state i to state j. Inventory at epoch i equals 0, so c_{ij} includes the holding cost $h_i(0)$. Exactly d_{ij} units are produced during period i, so c_{ij} includes the production cost $c_i(d_{ij})$. Exactly d_{kj} units of inventory remain at any epoch k between epochs i and j, so c_{ij} includes the holding cost

$h_k(d_{kj})$ for epoch k. This accounts for all components of c_{ij} and leads to the equation:[1]

(5-11) $$c_{ij} = h_i(0) + c_i(d_{ij}) + \sum_{k=i+1}^{j-1} h_k(d_{kj})$$

The cost of a transition from state i to state j is c_{ij}. With $f(N+1) = 0$, one gets the functional equation

(5-12) $$f(i) = \min_{j \mid i < j \leq N+1} \{c_{ij} + f(j)\}, \qquad i = 1, \ldots, N$$

Tables of d_{ij} and of c_{ij} can be built with work proportional to N^2, and functional equation (5-12) can be solved by reaching or by recursive fixing. Either way, the computational effort is proportional to N^2.

The most interesting feature of the concave-cost case is, perhaps, the structure of the optimal policy. If production occurs during period i, it should be equal to the total demand in periods i through $j-1$, for some $j > i$.

> *Remark:* This development, including Theorem 1 and functional equation (5-12), applies to the case in which x_n is allowed to vary continuously, provided that the functions c_n and h_n are concave on the set of nonnegative numbers.

BACKLOGGING

Until now, demand during a period has been satisfied by production during that period or during earlier periods. Demand is now allowed to accumulate and be satisfied by production during subsequent periods. This is known as *backlogging* demand. The only effect on Program 1 of allowing backlogging is to delete the requirement that the variables I_2 through I_N be nonnegative. Equation (5-5) still identifies I_n as the total production during periods 1 through $n - 1$, less the total demand during those periods. When I_n is nonnegative, it still represents an inventory of I_n units. When I_n is negative, it now represents a *shortage* of $-I_n$ units of unfilled (backlogged) demand that must be satisfied by production during periods n through N.

When backlogging is allowed, h_n is called the *holding/shortage* cost function for period n. When I_n is nonnegative, $h_n(I_n)$ remains equal to the cost of having I_n units of inventory on hand at the start of period n. When I_n is negative, $h_n(I_n)$ becomes the cost of having a shortage of $-I_n$ units of unfilled demand on hand at the start of period n. The model with backlogging admits of a functional equation much like the one given in (5-6) and (5-8).

Exercise 1. Adapt the functional equation given by (5-6) and (5-8) to the variant of Program 1 in which backlogging is allowed.
[Hint: Which production levels are now feasible?]

---

[1] When j equals $i + 1$, the summation over k in (5-11) is over the empty set and is defined to be zero.

BACKLOGGING WITH CONCAVE COSTS

The holding costs and the shortage costs are included in h_n. The model is (still) said to have concave holding costs if, for each n, the function $h_n(x)$ is concave on J. The model is now said to have *concave shortage costs* if, for each n, the function $h_n(-x)$ is concave on J. In the case of concave holding and shortage costs, the function h_n can take the form depicted in Figure 5-1. Figure 5-1

FIGURE 5-1. Concave holding/shortage cost h_n

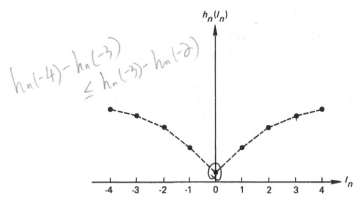

reflects the fact that the cost of increasing the shortage from 3 to 4 is no greater than the cost of increasing the shortage from 2 to 3. In other words, the model has concave holding and shortage costs if, for each n, the function h_n satisfies

(5-13) $$h_n(x + 2) - h_n(x + 1) \leq h_n(x + 1) - h_n(x)$$

for every integer x except, perhaps, for $x = -1$. The function h_n in Figure 5-1 violates (5-13) when $x = -1$. Use (5-13) with $x = -4$ to see that the cost of increasing the shortage from 3 to 4 [i.e., $h_n(-4) - h_n(-3)$] is no greater than the cost of increasing the shortage from 2 to 3 [i.e., $h_n(-3) - h_n(-2)$].

The function depicted in Figure 5-1 is not concave on the integers because (5-13) is violated when $x = -1$. Indeed, if (5-13) held for every n and x, one could show (see Problem 9) that it is optimal to concentrate all the production in one period.

When the production and holding shortage costs are concave, the solution to the production planning problem has the form given below.

THEOREM 2 (*backlogging with concave costs*).   Consider the variant of Program 1 that deletes the constraints $I_n \geq 0$ for $n = 2, \ldots, N$. Suppose that the production, holding, and shortage costs are concave. Then there exists at least one optimal production plan $(x_1, \ldots, x_N)$ having this property: If $x_m > 0$ and $x_n > 0$ for $m < n$, then $I_i = 0$ for at least one i satisfying $m < i \leq n$.

The proof of Theorem 2 is deferred to a later section, but an implication of it is drawn here. Let $x^* = (x_1, \ldots, x_N)$ be an optimal production plan of the type described in Theorem 2; that is, if production levels x_m and x_n are positive for periods $m < n$, then inventory I_i equals zero for some period i having $m < i \leq n$. Consider any period $i \leq N$ such that x^* has $I_i = 0$. (Since $I_1 = 0$, such an i exists.) Consider also the lowest-numbered $j > i$ such that x^* has $I_j = 0$. (Since $I_{N+1} = 0$, such a j exists.) Exactly d_{ij} units must be produced during periods i through $j - 1$. We argue by contradiction that production of these d_{ij} units is concentrated in one period k, where $i \leq k < j$. Suppose not: that is, that x^* splits this production between periods k and $l > k$. Inventory I_p must be zero for some period p having $k < p \leq l < j$. The minimality of j precludes this. Hence, production of these d_{ij} units is not split. For $i \leq k < j$, let $c_{ij}(k)$ denote the total cost incurred during periods i through $j - 1$ if the total demand d_{ij} occurring during these periods is satisfied by production produced in period k. One has

$$c_{ij}(k) = h_i(0) + c_k(d_{ij}) + \sum_{n=i+1}^{k} h_n(-d_{in}) + \sum_{n=k+1}^{j-1} h_n(d_{nj})$$

Exercise 2. Account for the components of $c_{ij}(k)$.

In a dynamic programming formulation, state i again denotes the situation of having a zero inventory on hand at epoch i. Transition occurs from state i to state j if it is decided to produce d_{ij} units during some intermediate period k. The cheapest transition from state i to state j costs c_{ij}^*, where

(5-14) $$c_{ij}^* = \min_{k \mid i \leq k < j} \{c_{ij}(k)\}$$

As usual, $f(i)$ is defined, for $i = 1, \ldots, N$, by

$$f(i) = \text{minimal cost of satisfying demands during periods } i \text{ through } N \text{ if } I_i = 0$$

With $f(N + 1) = 0$, one gets the functional equation

(5-15) $$f(i) = \min_{j \mid i < j \leq N+1} \{c_{ij}^* + f(j)\}, \qquad i = 1, \ldots, N$$

Functional equation (5-15) is similar to the functional equation for the concave-cost case without backlogging. The difference is that c_{ij}^* replaces c_{ij}. Once a table of c_{ij}^* is built, (5-15) can be solved with work proportional to N^2, just as in the case of no backlogging. However, the work needed to build a table of c_{ij}^* from (5-14) is proportional to N^3, not N^2. Problem 6 shows that in the case of linear production costs the computation can be reorganized to yield a solution to (5-15) with total effort proportional to N^2.

CONVEX COSTS

With $J = \{0, 1, \ldots\}$, the real-valued function g is called *convex on J* if

$$g(x + 2) - g(x + 1) \geq g(x + 1) - g(x), \qquad \text{all } x \text{ in } J$$

Interpret $g(x)$ as the cost of producing x items. If g is convex, the incremental cost of producing item $x + 2$ is no smaller than the incremental cost of producing item $x + 1$. This reflects *increasing marginal costs* or *diseconomies of scale*, these terms being synonymous with convex costs. The model is said to have *convex production and holding costs* if, for each n, the functions c_n and h_n are convex on J. In a later section, we shall analyze a convex-cost model that is more general than Program 1, and we shall prove the following theorem.

THEOREM 3 (*convex costs*). Consider Program 1 in the case of convex production and holding costs. Let $x = (x_1, \ldots, x_N)$ be an optimal production plan for vector $(d_1, \ldots, d_N)$ of demands. If one of these demands is increased by 1 unit, it is optimal to increase one of these production levels by 1 unit (although not, necessarily, in the same period).

A marginal analysis procedure arises directly from Theorem 3. Fill demands 1 unit at a time, at each iteration producing in a period whose marginal contribution to cost is minimal. This is reminiscent of the marginal analysis procedure in Chapter 3, which is as it should be, as we are dealing with a generalization of that model.

Exercise 3. Show that Program 4 of Chapter 3 is a special case of Program 1 of this chapter.
[Hint: Set $d_n = 0$ for $n < N$.]

PRODUCTION CONTROL AND NETWORK FLOW

Theorems 1 and 2 are proved here. As a first step, Program 1 is recast as a network flow problem. The "flow" occurs along arcs, and flow into each node must equal flow out of that node. More specifically, each arc in the network corresponds to a variable, and each node corresponds to an equation. Figure 5-2 contains a network flow representation of Program 1. The variables x_n and I_n correspond to arcs, and the arrows indicate the direction in which flow is allowed. For $n = 1, \ldots, N$, node n arises from (5-2); flow into node n equals $I_n + x_n$, and flow out of node n equals $d_n + I_{n+1}$. The flow of $D = d_1 + \ldots + d_N$ into node 0 reflects the fact that total production $x_1 + \ldots + x_N$ equals total demand D. Fixed flows, such as d_n and D, are connected to only one node. Variable flows, such as I_n and x_n, flow from one node to another. The requirement in Program 1 that I_1 and I_{N+1} be zero is represented by excluding these arcs from Figure 5-2. The costs of flows I_n and x_n are, respectively, $h_n(I_n)$ and $c_n(x_n)$.

The network flow model in Figure 5-2 captures all aspects of Program 1, except (check this) the requirement that each x_n be integer-valued. We shall call a flow *feasible* if its variables satisfy the constraints, including integrality,

FIGURE 5-2. Network flow representation of Program 1

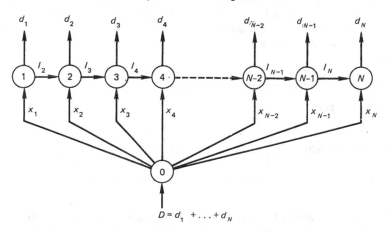

$$D = d_1 + \ldots + d_N$$

of Program 1. An *optimal flow* is then any feasible flow the sum of whose costs is minimal.

Loopless Flows

Consider a feasible flow. Call arc (i, j) *active* if flow along it is not zero. A *loop* is said to exist if one can start at a node and return to it by traversing a sequence of *distinct* active arcs, not necessarily in the direction of the arrows. Suppose, for instance, that x_3, I_4, and x_4 are all positive. Then the node sequence $(0, 3, 4, 0)$ prescribes a loop, as it is possible to traverse active arc $(0, 3)$, then active arc $(3, 4)$, and then active arc $(0, 4)$ in reverse, thereby returning to node 0. Notice that $(0, 3, 0)$ does not prescribe a loop, as arc $(0, 3)$ cannot be traversed twice.

THEOREM 4. A feasible flow for Program 1 has no loops if and only if it satisfies (5-16).

(5-16) $$I_j x_j = 0, \qquad j = 2, \ldots, N$$

Proof. Note that every loop in Figure 5-2 is of the form $(0, i, i + 1, \ldots, j, 0)$, where $1 \leq i < j \leq N$. Such a loop requires that x_j and I_j be positive, which violates (5-16). Now consider a feasible flow that violates (5-16). It has $I_j x_j > 0$ for some $j > 1$. Inventory I_j remains from production during some period $i < j$, and a loop $(0, i, \ldots, j, 0)$ exists. This proves Theorem 4. ∎

> *Remark:* Theorem 4 lets us sketch, for the *cognoscenti*, a sophisticated proof of Theorem 1. In Program 1, a concave function is being minimized over a compact convex set, and at least one optimal solution is an extreme point. Loops are defined so that a flow is an extreme point if and only if it contains no loops. Hence, Theorem 4 shows that (5-16) is satisfied by every extreme point and, in particular, by an optimal extreme point. This proves Theorem 1.

Forward differences are used in the less sophisticated proof of Theorem 1 that follows. Let g be a function whose domain is the nonnegative integers. Define its first forward difference Δg and its second forward difference $\Delta^2 g$ on the nonnegative integers by

(5-17) $\Delta g(x) = g(x + 1) - g(x),$ $x = 0, 1, \ldots$

(5-18) $\Delta^2 g(x) = \Delta g(x + 1) - \Delta g(x),$ $x = 0, 1, \ldots$

Substitute to get

(5-19) $\Delta^2 g(x) = g(x + 2) - 2g(x + 1) + g(x),$ $x = 0, 1, \ldots$

Compare (5-9) with (5-17) to (5-19) to see that the following three statements are equivalent: g is concave on J; the first forward difference of g is nonincreasing; the second forward difference of g is nonpositive [i.e., $\Delta^2 g(x) \leq 0$ for $x = 0, 1, \ldots$].

Proof of Theorem 1.   There are finitely many feasible production plans for Program 1. Hence, there exists at least one optimal production plan for Program 1. Select an optimal production plan $\mathbf{x} = (x_1, \ldots, x_N)$ so that it and $\mathbf{I} = (I_1, \ldots, I_N)$, as determined by (5-5), minimize the sum, $\sum_{i=1}^{N} (x_i + I_i)$. (This is a tie-breaking rule.) Aiming for a contradiction, we assume that $\mathbf{x}$ and $\mathbf{I}$ violate (5-16). Theorem 4 shows that a loop exists; it is necessarily of the form $(0, i, \ldots, j, 0)$ for $1 \leq i < j \leq N$. All flows along this loop are positive integers. A feasible flow results if we increase x_i by 1, decrease x_j by 1 and, consequently, increase I_k by 1 for $i < k \leq j$. This perturbed flow cannot decrease cost below the optimum. So

$$c_i(x_i) + c_j(x_j) + \sum_{k=i+1}^{j} h_k(I_k) \leq c_i(x_i + 1) + c_j(x_j - 1) + \sum_{k=i+1}^{j} h_k(I_k + 1)$$
(5-20)

It is also feasible to perturb $\mathbf{x}$ by decreasing x_i by 1, increasing x_j by 1 and, consequently, decreasing I_k by 1 for $i < k \leq j$. This perturbed flow cannot decrease cost below the optimum, and so

$$c_i(x_i) + c_j(x_j) + \sum_{k=i+1}^{j} h_k(I_k) \leq c_i(x_i - 1) + c_j(x_j + 1) + \sum_{k=i+1}^{j} h_k(I_k - 1)$$
(5-21)

Add (5-20) to (5-21), and rearrange the sum as

(5-22) $0 \leq \Delta^2 c_i(x_i - 1) + \Delta^2 c_j(x_j - 1) + \sum_{k=i+1}^{j} \Delta^2 h_k(I_k - 1)$

Second forward differences of concave functions are nonpositive. So the right-hand side of (5-22) is the sum of nonpositive quantities. So inequality (5-22) must hold as an equation. So inequalities (5-20) and (5-21), of which (5-22) is the sum, must also hold as equations. So both perturbations of $(\mathbf{x}, \mathbf{I})$ are optimal flows. The perturbation that decreases x_i by 1, increases x_j by 1, and decreases I_k by 1 for $i < k \leq j$ decreases the sum of the flows on the arcs by $(j - i)$.

This contradicts the supposed minimality of that sum, which completes the proof. ■

Backlogging with Concave Costs

Consider the variant of Program 1 in which backlogging is allowed. The effect of this is to delete the nonnegativity constraint on I_n, for $n = 2, \ldots, N$. The network flow representation in Figure 5-2 remains applicable, provided one understands that, for $n = 2, \ldots, N$, the flow I_n along arc $(n - 1, n)$ can be negative. Call arc (i, j) *active* if flow along it is nonzero. Flows on active arcs can now be negative. A loop has the same definition as before.

Proof of Theorem 2. Select an optimal production plan $\mathbf{x} = (x_1, \ldots, x_N)$ so that it and $\mathbf{I} = (I_1, \ldots, I_N)$, as determined by (5-5), minimize the sum $\sum_{i=1}^{N} (x_i + I_i)$. Aiming for a contradiction, assume that flow $(\mathbf{x}, \mathbf{I})$ contains a loop. Proceed exactly as in the proof of Theorem 1 to contradict the supposed minimality of the foregoing sum. Hence, $(\mathbf{x}, \mathbf{I})$ contains no loop. Consequently, if x_m and x_n are positive for some $m < n$, then $I_i = 0$ for some i such that $m < i \leq n$. This completes the proof of Theorem 2. ■

> *Remark:* The cognoscenti will recognize an active arc as one whose flow is in the interior of a region of concavity. As before, no flow whose active arcs contain a loop can be an extreme point; hence, at least one optimal flow is loopless.

NETWORK FLOW WITH CONVEX COSTS*

This section reflects the generic condition that convex-cost models are more tractable than concave-cost models. Rather than proving Theorem 3 directly, we obtain it as a biproduct of more general results. Instead of studying Program 1, we analyze the general network flow model that is now introduced. The nodes in this network form the finite set S. The arcs of this network form the subset A of $S \times S$. Assume that this network has no self-loops [i.e., no arcs of the form (i, i)]. For each node i, define sets E_i and T_i of nodes by

$$E_i = \{j \mid (i, j) \in A\}, \qquad T_i = \{j \mid (j, i) \in A\}$$

Hence, each arc $(i, j) \in A$ has $j \in E_i$ and $i \in T_j$.

The data in the network flow model are the network (S, A), the integers b_i and U_{ij}, and the functions c_{ij} that appear in

Program 2. Minimize $\{\sum_{(i, j) \in A} c_{ij}(x_{ij})\}$, subject to the constraints

(5-23)	$0 \leq x_{ij} \leq U_{ij}$,	all $(i, j) \in A$
(5-24)	x_{ij} integer,	all $(i, j) \in A$
(5-25)	$\sum_{j \in E_i} x_{ij} - \sum_{j \in T_i} x_{ji} = b_i$,	all $i \in S$,

---

* This section is self-contained, but it uses ideas that are easier to grasp if one has a thorough knowledge of linear programming and of the supplement on convexity.

Interpret x_{ij} as the flow along arc (i, j). Interpret U_{ij} as the upper bound on this flow. Possibly, $U_{ij} = +\infty$. Interpret $c_{ij}(x_{ij})$ as the cost of sending x_{ij} units of flow along arc (i, j). Interpret the (possibly negative) integer b_i as the net flow into node i. Sum (5-25) over all nodes i to obtain $0 = \sum_{i \in S} b_i$. Hence, Program 2 cannot have a feasible solution unless the constants b_i sum to zero. A five-node example of Program 2 is provided as Figure 5-3. The transportation network in Figure 5-3 makes it clear that Program 2 is not restricted to production control problems. Program 2 is a general model of flows in networks.

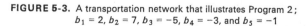

FIGURE 5-3. A transportation network that illustrates Program 2;
$b_1 = 2$, $b_2 = 7$, $b_3 = -5$, $b_4 = -3$, and $b_5 = -1$

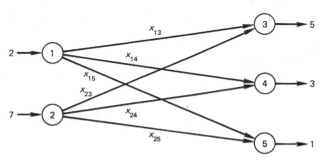

The network in Figure 5-2 is easy to cast in the form given in Program 2; set $S = \{0, 1, \ldots, N\}$, set $b_i = -d_i$ for $i = 1, \ldots, N$, set $b_0 = D$, and so on.

The network flow model in Program 2 is analyzed under the assumption that each cost function c_{ij} is convex. This includes, of course, the case of linear costs. For each set $\mathbf{x} = \{x_{ij} | (i, j) \in A\}$ of flow variables, write $c(\mathbf{x}) = \sum_{(i, j) \in A} c_{ij}(x_{ij})$. Call $\mathbf{x}$ a *feasible flow* if it satisfies (5-23) to (5-25). Call a feasible flow $\mathbf{x}$ an *optimal flow* if $c(\mathbf{x}) \le c(\mathbf{y})$ for every other feasible flow $\mathbf{y}$.

An optimal flow is identified by examining perturbations of feasible flows that allow increases in the flow along certain arcs and decreases in the flow along others. Such a perturbation is described as an analog of a "path" in which arcs can be traversed in either the forward (increased flow) direction or in the reverse (decreased flow) direction. Specifically, a *chain* is an alternating sequence $[i_0, a_1, i_1, \ldots, a_p, i_p]$ of nodes i_0 through i_p and arcs a_1 through $a_p (p \ge 1)$ with the property that each arc a_n equals (i_{n-1}, i_n) or (i_n, i_{n-1}). Arc a_n in this chain is called a *forward* arc if $a_n = (i_{n-1}, i_n)$; it is a *reverse* arc if $a_n = (i_n, i_{n-1})$. This chain is called *arc-disjoint* if no arc repeats.

The network in Figure 5-3 is used to illustrate these definitions. Note that $[1, (1, 3), 3, (1, 3), 1]$ is a chain, but not an arc-disjoint chain, because arc $(1, 3)$ repeats. Note that $[1, (1, 3), 3, (2, 3), 2, (2, 4), 4]$ is an arc-disjoint chain; arcs $(1, 3)$ and $(2, 4)$ are forward arcs, and arc $(2, 3)$ is a reverse arc in this chain.

An arc-disjoint chain has *chain flow* $\mathbf{z} = \{z_{ij} | (i, j) \in A\}$ specified as follows: $z_{ij} = 1$ if (i, j) is a forward arc in the chain; $z_{ij} = -1$ if (i, j) is a reverse arc in the chain; and $z_{ij} = 0$ if (i, j) is not an arc in the chain. The arc-disjoint chain in the preceding paragraph has chain flow $\mathbf{z}$ with $z_{13} = z_{24} = 1$, $z_{23} = -1$, and $z_{15} = z_{14} = z_{25} = 0$.

Let $\mathbf{x} = \{x_{ij} | (i, j) \in A\}$ be any set of flow variables, and let $\mathbf{z}$ be any chain flow. Call $\mathbf{z}$ a *feasible perturbation* of $\mathbf{x}$ if $\mathbf{x} + \mathbf{z}$ satisfies the constraints of Program 2. Also define $m(\mathbf{z}; \mathbf{x})$ by

$$m(\mathbf{z}; \mathbf{x}) = c(\mathbf{x} + \mathbf{z}) - c(\mathbf{x}) = \sum_{(i,j) \in A} \{c_{ij}(x_{ij} + z_{ij}) - c_{ij}(x_{ij})\}$$

In the exercise that follows, you are asked to identity $m(\mathbf{z}; \mathbf{x})$ as the length of a directed path.

Exercise 4. The network in Figure 5-3 has feasible flow $\mathbf{x}$ with $x_{13} = x_{14} = x_{25} = 1$, $x_{23} = 4$, $x_{24} = 2$, and $x_{15} = 0$. Consider the arc disjoint chain $[i_0, a_1, i_1, \ldots, i_p]$ given as $[1, (1, 3), 3, (2, 3), 2, (2, 4), 4, (1, 4), 1]$. Write the chain flow $\mathbf{z}$ for this chain. With $c_{ij}(x_{ij})$ as the cost of flow x_{ij} on arc (i, j), define a set of arcs and arc lengths so that $m(\mathbf{z}; \mathbf{x})$ is the length of the directed path $(i_0, i_1, \ldots, i_p)$.

Call an arc-disjoint chain $[i_0, a_1, \ldots, i_p]$ a *simple circuit* if nodes i_0 through i_{p-1} are distinct and if node $i_p = i_0$. A simple circuit flow $\mathbf{z}$ has

$$\sum_{j \in E_i} z_{ij} - \sum_{j \in T_i} z_{ji} = 0, \quad \text{all } i \in S$$

Note that chain $[1, (1, 3), 3, (1, 3), 1]$ is not a simple circuit because it is not arc-disjoint. A simple circuit flow $\mathbf{z}$, such as the one in Exercise 4, conserves flow at all nodes. Simple circuits play a central role in the analysis of Program 2 that follows.

THEOREM 5. Suppose, in Program 2, that each function c_{ij} is convex.

(a) Consider any $\mathbf{x}$ and $\mathbf{y} \neq \mathbf{x}$ that satisfy (5-24) and (5-25). There exist a positive integer q and simple circuit flows $\mathbf{z}^1, \ldots, \mathbf{z}^q$ such that

(5-26) $$\mathbf{y} - \mathbf{x} = \mathbf{z}^1 + \ldots + \mathbf{z}^q$$

(5-27) $$(y_{ij} - x_{ij})z_{ij}^r = 0, \quad \text{all } i, j, r$$

(5-28) $$c(\mathbf{y}) - c(\mathbf{x}) \geq m(\mathbf{z}^1; \mathbf{x}) + \ldots + m(\mathbf{z}^q; \mathbf{x})$$

(b) A feasible flow $\mathbf{x}$ is optimal if and only if $m(\mathbf{z}; \mathbf{x}) \geq 0$ for every simple circuit flow $\mathbf{z}$ that is a feasible perturbation of $\mathbf{x}$.

Remark: The proof of Theorem 5 is intricate. It can be skipped without loss of continuity. The cognoscenti will recognize that a convex function is being minimized over a convex set, so that a "local" minimum is a global minimum. In this context, Theorem 5 supplies conditions that suffice for a local minimum.

Proof. Let $\mathbf{w} = \mathbf{y} - \mathbf{x}$ and note that

(5-29) $$\sum_{j \in E_i} w_{ij} - \sum_{j \in T_i} w_{ji} = 0, \quad \text{all } i \in S$$

Let $B = \{(i, j) \mid w_{ij} > 0 \text{ and/or } w_{ji} < 0\}$. Start with an arc (i, j) in B, and trace out a (directed) path in the directed network (S, B) until the first time a node repeats. The intervening arcs identify a (directed) cycle $(i_0, \ldots, i_p)$ in (S, B) and a simple circuit $[i_0, a_t, \ldots, a_p, i_p]$ of (S, A) with forward (respectively, reverse) arc (i, j) having $w_{ij} > 0$ (respectively, $w_{ji} < 0$). This simple circuit identifies a simple circuit flow $\mathbf{z}^1$ that satisfies $z_{ij}^1 w_{ij} \geq 0$ for all (i, j). If $\mathbf{w} = \mathbf{z}^1$, stop. Otherwise, replace $\mathbf{w}$ by $\mathbf{w} - \mathbf{z}^1$, redefine B, and iterate, calling successive simple circuit flows $\mathbf{z}^2$, $\mathbf{z}^3$, and so on. Each iteration reduces $\sum_{(i, j)} |w_{ij}|$ by a positive integer while satisfying (5-27) and preserving (5-29). Consequently, this process terminates in finitely many iterations with a solution to (5-26) and (5-27).

Expression (5-28) is now verified. Set $\mathbf{u}^0 = \mathbf{x}$ and $\mathbf{u}^r = \mathbf{x} + \mathbf{z}^1 + \ldots + \mathbf{z}^r$ for $r = 1, \ldots, q$. Consider first the case $z_{ij}^r = +1$. In this case, (5-27) assures that $x_{ij} \leq u_{ij}^{r-1} < u_{ij}^r = u_{ij}^{r-1} + z_{ij}^r$.

The forward difference of a convex function is nondecreasing; hence,

$$(5\text{-}30) \qquad c_{ij}(x_{ij} + z_{ij}^r) - c_{ij}(x_{ij}) \leq c_{ij}(u_{ij}^r) - c_{ij}(u_{ij}^{r-1})$$

Now consider the case $z_{ij}^r = -1$. In this case (5-27) assures that $x_{ij} \geq u_{ij}^{r-1} > u_{ij}^r = u_{ij}^{r-1} + z_{ij}^r$. The forward difference of a convex function is nondecreasing, which again yields (5-30). Hence, (5-30) holds for $z_{ij}^r \in \{+1, -1\}$. Sum to get

$$(5\text{-}31) \qquad c(\mathbf{x} + \mathbf{z}^r) - c(\mathbf{x}) \leq c(\mathbf{u}^r) - c(\mathbf{u}^{r-1})$$

Sum (5-31) over r to verify (5-28), which completes the proof of part (a).

Suppose $\mathbf{x}$ is a feasible flow and $m(\mathbf{z}; \mathbf{x}) < 0$ for some simple circuit flow $\mathbf{z}$ that is a feasible perturbation of $\mathbf{x}$. Then $\mathbf{x} + \mathbf{z}$ is a feasible flow, and $c(\mathbf{x} + \mathbf{z}) = c(\mathbf{x}) + m(\mathbf{z}; \mathbf{x}) < c(\mathbf{x})$. Hence, $\mathbf{x}$ is not optimal. Now suppose that $\mathbf{x}$ is feasible and $m(\mathbf{z}; \mathbf{x}) \geq 0$ for every simple circuit flow $\mathbf{z}$ that is a feasible perturbation of $\mathbf{x}$. Consider any feasible flow $\mathbf{y}$. Simple circuit flows $\mathbf{z}^1, \ldots, \mathbf{z}^q$ exist that satisfy (5-26) to (5-28). Moreover, (5-27) assures that $\mathbf{z}^r$ is a feasible perturbation of $\mathbf{x}$, for $r = 1, \ldots, q$. Hence, $m(\mathbf{z}^r; \mathbf{x}) \geq 0$, and (5-28) assures that $c(\mathbf{y}) \geq c(\mathbf{x})$. So $\mathbf{x}$ is optimal, which completes the proof. ∎

Simple Cycles and Optimal Flows

Theorem 5 demonstrates that an optimal flow can be found by starting with a feasible flow $\mathbf{x}$ and searching for a simple-circuit flow $\mathbf{z}$ such that $\mathbf{x} + \mathbf{z}$ is feasible and $m(\mathbf{z}; \mathbf{x}) < 0$. If no such $\mathbf{z}$ exists, $\mathbf{x}$ is an optimal flow. The search for such a $\mathbf{z}$ amounts to the search for a cycle whose length is negative in the directed network described in Exercise 5, below.

Exercise 5. Let $\mathbf{x}$ be a feasible flow for Program 2. Show that there exists a simple-circuit flow $\mathbf{z}$ such that $\mathbf{x} + \mathbf{z}$ is a feasible flow and such that $m(\mathbf{z}; \mathbf{x}) < 0$ if and only if a cycle exists whose length is negative in the directed network (S, B) whose arc set B is as follows. If $x_{ij} < U_{ij}$, include in B an arc (i, j) whose length is $c_{ij}(x_{ij} + 1) - c_{ij}(x_{ij})$. If $x_{ij} > 0$, include in B an arc (j, i) whose length is $c_{ij}(x_{ij} - 1) - c_{ij}(x_{ij})$. If this creates a pair of parallel arcs, delete the longer.

Cycles whose lengths are negative can be detected efficiently by linear programming or by the label-correcting method given in Problem 21 of Chapter 2.

Shortest Paths and Optimal Flows

The way in which perturbing the b_i's changes the optimal flow for Program 2 is now investigated. Let Program 2_{st} denote the perturbation of Program 2 in which b_s is increased by 1, b_t is decreased by 1, and all other data are left unchanged. We shall see how to obtain the solution to Program 2_{st} from the solution to Program 2 by solving a shortest-path problem. Call a chain $[i_0, a_1, \ldots, i_p]$ an *elementary chain from* i_0 *to* i_p if nodes i_0 through i_p are disjoint. An elementary chain is necessarily arc disjoint. If z is an elementary chain flow, call $m(z; x)$ the *length* of its elementary chain. [This suppresses dependence of $m(z; x)$ on x.]

The network in Figure 5-3 is used to illustrate these definitions. Program 2_{14} has b_1 increased from 2 to 3 and b_4 decreased from -3 to -4. Two of the elementary chains from node 1 to node 4 are $[1, (1, 4), 4]$ and $[1, (1, 3), 3, (2, 3), 2, (2, 4), 4]$. With respect to a set x of flow variables, the latter chain has length $[c_{13}(x_{13} + 1) - c_{13}(x_{13}) + c_{23}(x_{23} - 1) - c_{23}(x_{23}) + c_{24}(x_{24} + 1) - c_{24}(x_{24})]$.

Let x be a feasible flow for Program 2, and let z be the chain flow for an elementary chain from s to t. Note that $x + z$ satisfies constraints (5-24) and (5-25) of Program 2_{st}. If

$$(5\text{-}32) \qquad 0 \le x_{ij} + z_{ij} \le U_{ij}, \qquad \text{all } (i, j) \in A$$

then $x + z$ is feasible for Program 2_{st}.

THEOREM 6 (*perturbation of Program 2*). Let x be an optimal flow for Program 2, and assume that each function c_{ij} is convex on $J = \{0, 1, \ldots\}$.

  a. Suppose that z is the chain flow for the shortest elementary chain from s to t such that $(x + z)$ satisfies (5-32). Then $(x + z)$ is an optimal flow for Program 2_{st}.

  b. Suppose that no elementary chain from s to t has chain flow z such that $(x + z)$ satisfies (5-32). Then Program 2_{st} has no feasible flow.

  Remark: The proof of Theorem 6 is intricate, and it can be skipped without loss of continuity.

  Proof. It suffices for parts (a) and (b) to consider any feasible flow y for Program 2_{st} and show that $c(y) \ge c(x + z)$, where z is the chain flow for some elementary chain from s to t having $0 \le x_{ij} + z_{ij} \le U_{ij}$ for all $(i, j) \in A$. Imagine that network (S, A) is augmented by inclusion of the node α and the new arcs (t, α) and (α, s) with the new data $U_{t\alpha} = U_{\alpha s} = 1$, $b_\alpha = 0$, and $c_{t\alpha}(\cdot) = c_{\alpha s}(\cdot) \equiv 0$. In this imagined network, set $x_{t\alpha} = x_{\alpha s} = 0$ and $y_{t\alpha} = y_{\alpha s} = 1$.

Apply part (a) of Theorem 5 to the imagined network to get the following result in terms of the actual $\mathbf{x}$ and $\mathbf{y}$: $\mathbf{y} - \mathbf{x} = \mathbf{z}^1 + \ldots + \mathbf{z}^q$, where $\mathbf{z}^1$ through $\mathbf{z}^q$ satisfy (5-27), (5-28), and (5-32), where $\mathbf{z}^1$ through $\mathbf{z}^{q-1}$ are simple-circuit flows and where $\mathbf{z}^q$ is the chain flow for an elementary chain from s to t. Optimality of $\mathbf{x}$ gives $m(\mathbf{z}^i; \mathbf{x}) \geq 0$ for $i < q$, hence, (5-28) gives $c(\mathbf{y}) \geq c(\mathbf{x}) + m(\mathbf{z}^q; \mathbf{x}) = c(\mathbf{x} + \mathbf{z}^q)$. Since $\mathbf{x} + \mathbf{z}^q$ is feasible for Program 2_{st}, the proof is complete. ■

Theorem 6 shows that perturbing b_s and b_t by 1 and -1, respectively, causes the optimal flow $\mathbf{x}$ to be perturbed to $\mathbf{x} + \mathbf{z}$, where $\mathbf{z}$ is the chain flow for a shortest elementary chain from node s to node t, subject to feasibility constraint (5-32). A similar result is obtained in Problem 10 for the perturbation of Program 2 in which U_{ij} is increased by one.

In the exercise that follows, you are asked to use the directed network in Exercise 5 to find the shortest elementary chain from s to t whose chain flow $\mathbf{z}$ satisfies (5-32).

Exercise 6. Let $\mathbf{x}$ be an optimal flow for Program 2, and let $(i_0, i_1, \ldots, i_p)$ be the shortest path from s (so $i_0 = s$) to t (so $i_p = t$) in the directed network in Exercise 5. Show how to determine an elementary chain flow $\mathbf{z}$ so that $\mathbf{x} + \mathbf{z}$ is an optimal flow for Program 2_{st}.

Proof of Theorem 3. Let $\mathbf{x}$ be an optimal flow for Program 1. Apply Theorem 6 to the network in Figure 5-2 to see that increasing d_j and D by 1 augments $\mathbf{x}$ by an elementary chain flow $\mathbf{z}$ that has $z_{0n} = 1$ for period n, the one period in which production is increased. ■

Theorems 5 and 6 underlie the following method [see Hu (1966)] for solving Program 2. Let $P = \{i \,|\, b_i > 0\}$ and $N = \{i \,|\, b_i < 0\}$. Take $x_{ij} \equiv 0$ as the initial feasible flow for Program 2 with $b_i \equiv 0$. Search for cycles of negative length until no more can be found. This gives an optimal flow for Program 2 with $b_i \equiv 0$. Then, repeatedly: Seek the shortest of the paths from nodes i in P to nodes j in N in the network in Exercise 5; replace $\mathbf{x}$ by $\mathbf{x} + \mathbf{z}$ where $\mathbf{z}$ is this path's elementary chain flow; and decrement b_i (increment b_j) by 1, removing i from P (j from N) when b_i (b_j) becomes 0. The number of shortest-path problems encountered by this method equals $\sum_{i \in P} b_i$, and each shortest-path problem is solved with work proportional to $|S|^2$, at worst. If, at some iteration, no path exists from P to N, Program 2 has no feasible flow.

Exercise 7. What happens to the computational procedure described in the preceding paragraph when Program 2 is unbounded (i.e., when its objective can approach $-\infty$)?

Exercise 8. Adapt the network in Exericse 5 to allow efficient computation of the shortest path from any node i in P to any node j in N.
[Hint: Add two "metanodes."]

SUMMARY

This chapter examines a model of production control. Particular attention is paid to the cases of concave costs and of convex costs. Analyses of both cases lead to the study of network flow models. In the concave-cost case, solutions are obtained by analyzing the structure of the networks encountered. In the convex-cost case, solutions are obtained as biproducts of general results for network flow models having convex costs. The following starred section sketches a multicommodity analogue of the convex-cost network flow model.

MULTICOMMODITY NETWORK FLOW WITH CONVEX COSTS*

Consider a directed network (S, A) where, for simplicity, $A = S \times S$. Let M be a finite set whose elements are called *commodities*. Interpret x_{ijk} as the flow of commodity k on arc (i, j). With $\mathbf{x} = \{x_{ijk} | (i, j) \in A, k \in M\}$, interpret $c(\mathbf{x})$ as the cost of flow $\mathbf{x}$. Consider the variant of Program 2 that follows.

Program 3. Minimize $c(\mathbf{x})$ subject to the constraints

(5-33) $$0 \leq x_{ijk} \leq U_{ijk}, \qquad \text{all } (i, j) \in A, k \in M$$

(5-34) $$\sum_{j \neq i} (x_{ijk} - x_{jik}) = b_{ik}, \qquad \text{all } i \in S, k \in M$$

Recognize Program 3 as $|M|$ one-commodity flow problems that are linked *solely* through the objective $c(\mathbf{x})$. Program 3 is studied here under the assumption that $c(\cdot)$ is convex *and* differentiable. Assume, in addition, that $U_{iik} \equiv 0$ (i.e., that the network has no self-loops). Each flow problem in Program 3 is like the one in Program 2 except that the integrality constraints are omitted and the objective $c(\mathbf{x})$ is not separable.

Call flow $\mathbf{x}$ *feasible* if it satisfies (5-33) and (5-34), and call a feasible flow $\mathbf{x}$ *optimal* if $c(\mathbf{x}) \leq c(\mathbf{y})$ for every feasible flow $\mathbf{y}$. A *simple circuit* for commodity k is defined exactly as before, and its *simple-circuit flow* $\mathbf{z}$ is defined exactly as before, but with $z_{ijl} = 0$ whenever $l \neq k$. Also, call simple circuit flow $\mathbf{z}$ a *feasible perturbation* of $\mathbf{x}$ if $\mathbf{x} + \epsilon\mathbf{z}$ is a feasible flow for all sufficiently small positive ϵ. In addition, redefine $m(\mathbf{z}; \mathbf{x})$ by

(5-35) $$m(\mathbf{z}; \mathbf{x}) = \lim_{\epsilon \to 0} \frac{c(\mathbf{x} + \epsilon\mathbf{z}) - c(\mathbf{x})}{\epsilon}$$

One gets the following analogue of Theorem 5.

---

* This section concerns a special topic that is related to Theorem 5, but not to other aspects of this volume. It will be easier to understand if the student has some knowledge of linear programming.

THEOREM 7.   Suppose, in Program 3, that c is convex and differentiable.

(a) Consider $\mathbf{x}$ and $\mathbf{y} \neq \mathbf{x}$ that satisfy (5-33) and (5-34). There exist a positive integer q, simple-circuit flows $\mathbf{z}^1, \ldots, \mathbf{z}^q$ and positive scalars $\alpha^1, \ldots, \alpha^q$ such that

(5-36) $$\mathbf{y} - \mathbf{x} = \alpha^1 \mathbf{z}^1 + \ldots + \alpha^q \mathbf{z}^q$$

(5-37) $$(y_{ijk} - x_{ijk})z^r_{ijk} \geq 0, \qquad \text{all } i, j, k, r$$

(5-38) $$c(\mathbf{y}) - c(\mathbf{x}) \geq \alpha^1 m(\mathbf{z}^1; \mathbf{x}) + \ldots + \alpha^q m(\mathbf{z}^q; \mathbf{x})$$

(b) A feasible flow $\mathbf{x}$ is optimal if and only if $m(\mathbf{z}; \mathbf{x}) \geq 0$ for every simple circuit flow $\mathbf{z}$ that is a feasible perturbation of $\mathbf{x}$.   ■

Except for (5-38), the proof of Theorem 7 follows the same pattern as for Theorem 5. Proof of (5-38), which we omit, relies on the convexity *and* differentiability of $c(\cdot)$. To see what goes wrong when c is convex and not differentiable, look at Problem 10 of Supplement 2.

Program 3 is often used to model traffic flow in road networks. In this application, the nodes model road junctions, arc (i, j) models a road from junction i to junction j, and U_{ij} is the flow capacity of road (i, j). Commodity k corresponds to a group of travelers, and x_{ijk} is the number of type-k travelers who used road (i, j). Hence, the total number of travelers on road (i, j) equals $t_{ij} = \sum_k x_{ijk}$, and the objective of Program 3 takes the form

(5-39) $$c(x) = \sum_{(i, j) \in A} g_{ij}(t_{ij})$$

In the *monopolistic* version of this model, Program 3 is used to allocate travelers to roads so as to minimize the total travel time of the population. In the *equilibrium point* version, Program 3 is used to show how individual travelers select roads so as to minimize their own travel times. In the monopolistic version, $g_{ij}(t_{ij})$ is taken as the total travel time of the t_{ij} users of road (i, j). In the equilibrium point version, $g_{ij}(t_{ij})$ is taken as the travel time of each of the t_{ij} users of road (i, j). In either version, one is led to functions g_{ij} that are convex, differentiable, and nondecreasing.

BIBLIOGRAPHIC NOTES

The production control problem has received considerable attention in the literature. Dynamic programming formulations for the concave-cost case were due initially to Wagner and Whitin (1957) and, independently, to Manne (1958). Extensions to backlogging are due to Zangwill (1966, 1969) and to Manne and Veinott (1967). Of the later developments, we cite Fong and Rao (1973), and Erickson (1978). Theorem 3 was proved and generalized by Veinott (1964, 1973).

An extensive literature exists on network flow models with linear and with convex costs. The genesis of Theorem 5 is in Beale (1959). For related work on network flows, see Ford and Fulkerson (1962); Bradley, Brown, and Graves

(1977); Glover, Karney, and Klingman (1974); Klincewicz (1979); and Zadeh (1979).

Monopolistic and equilibrium-point objectives for the multicommodity network flow problem were proposed by Wardrop (1952), with subsequent contributions by Charnes and Cooper (1961), Dafermos and Sparrow (1969), Potts and Oliver (1972), Nguyen (1974), Florian and Nguyen (1976), and many others. Theorem 7 applies the *elementary vectors* in Rockafellar (1972); for general results and related literature, see his Theorem 22.6 and his notes (1972, p. 406).

PROBLEMS

1. (*Planning horizon*) Consider the convex-cost case of Program 1. Let $x = (x_1, \ldots, x_N)$ be an optimal production plan for vector $(d_1, \ldots, d_N)$ of demands. Show that vector $(d_1, \ldots, d_N, d_{N+1})$ of demands has optimal production plan $y = (y_1, \ldots, y_{N+1})$, where $y_i \geq x_i$ for $i = 1, \ldots, N$. (Hence, increasing the planning horizon does not decrease the optimal production levels.)

[Hint: Apply Theorem 3.]

State why this result need not hold if backlogging is allowed.

[Hint: Theorem 6 may be relevant.]

2. Sogrim liquor markets whiskey in qualities 1 through N, quality 1 being highest, quality 2 next highest, and so on. Sales orders have been received for d_i cases of whiskey of quality i, but customers will accept delivery of higher-quality whiskey than they ordered. If Sogrim bottles any whiskey of quality i, it incurs setup cost s_i as well as unit cost c_i per case bottled. Sogrim wishes to fill its orders at minimum total cost.

a. Show that Program 1 applies.

[Hint: Let I_n be the excess production of qualities 1 through $n - 1$, and define $h_n(I_n)$ correctly.]

b. Show that equation (5-12) specifies the optimal assortment of liquor to bottle.

3. (*Seasonal Sales*) Demands for a product during the months January through December are for 2, 2, 2, 2, 2, 2, 2, 2, 5, 7, 6, and 10 units, respectively. Regular-time production capacity is 3 units per month. Overtime capacity is 2 units per month. Regular-time and overtime production costs are $100 and $120 per unit, independent of the month. Inventory carrying costs are $3 per unit per month. Find an optimal production plan, and use Theorem 5 to verify that it is optimal.

4. Suppose that the production control model with no backlogging has linear costs and limits on production and on inventory: specifically, for each n,

$$c_n(x_n) = c_n x_n, \quad h_n(I_n) = h_n I_n, \quad x_n \leq P \text{ and } I_n \leq Q$$

n	1	2	3	4	5
X_n			~~3~~~~~2~~	~~$2/3/4$~~	~~$0/1/2$~~ $0/1/2$
I_n	2	1	~~2~~~~2~~	~~0~~	

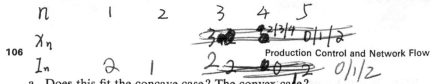

106

Production Control and Network Flow

a. Does this fit the concave case? The convex case?

b. Specialize one of the algorithms described in this chapter and apply it to the data; $N = 5$, $P = 4$, $Q = 2$, $h_n \equiv 1$, and

n	1	2	3	4	5
d_n	2	3	5	2	3
c_n	16	16	17	20	21

Surplus

5. [Manne and Veinott (1967)]. Demand for electricity equals D_n units during period n, where $D_1 \leq D_2 \leq \ldots \leq D_N$. Initial generating capacity is zero, and capacity at the start of each period must be sufficient to satisfy demand during that period. Let $c_n(x)$ denote the cost of adding x units of generating capacity at the start of period n. Once generating capacity is added, it remains in place for the remainder of the planning horizon. Assume that $c_n(x)$ is a concave function of x, reflecting economies of scale.

a. Formulate the problem of satisfying demand during periods 1 through N at minimal cost for *efficient* solution by dynamic programming.

b. Solve the variant of this problem in which electricity can also be purchased from external sources during any period or periods, assuming that the cost $h_n(y)$ of purchasing y units of electricity externally during period n has decreasing marginal cost in y.

6. [Zangwill (1969)]. Reconsider the production control problem with concave costs and backlogging. Notice that $c_{ij}(k) = A_{ik} + B_{kj} + c_k(d_{ij})$, where

$$C_k(d_{ij}) = M_k + C_k d_{ij}$$
$$= M_k + C_k(d_{ik} + d_{kj})$$

$$A_{ik} = h_i(0) + \sum_{n=i+1}^{k} h_n(-d_{in})$$

$$B_{kj} = \sum_{n=k+1}^{j-1} h_n(d_{nj})$$

a. Justify the functional equation

$$f_i = \min_{k \geq i} \{A_{ik} + g(i, k)\}$$
$$g(i, k) = \min_{j > k} \{B_{kj} + c_k(d_{ij}) + f_j\}$$

$O(N^3)$ k N^3 $\sqrt{i, j}$

b. Eliminate one state variable in part a for the special case. N^2

$$c_n(x) = \begin{cases} 0, & x = 0 \\ K_n + x \cdot L_n, & x > 0 \end{cases}$$

c. Argue that computation time in parts a and b are proportional to N^3 and N^2, respectively.

Note: The key idea used to analyze the concave-cost case of production control is that one need only examine flows whose active arcs form no loops, where an active arc is one whose flow is in the *interior of a region of concavity*. This idea is also a key to Problems 7 to 9.

7. Reconsider the production control model with concave costs. Allow back-logging, but for no more than two periods [i.e., $-(d_t + d_{t+1}) \leq I_{t+2}$].

    a. Redefine the active arcs so that the optimal flow is loopless.

    b. Write down an analogue of Theorem 2 (e.g., if $x_m > 0$ and $x_n > 0$ for some $m < n$, then . . .).

    c. Use part b to write an analogue of functional equation (5-14) and (5-15).

[**Hint:** Use two states for period i, one with $I_i = 0$, and the other with $I_i = -(d_{i-1} + d_{i-2})$.]

8. [Love (1973)]. Reconsider the production control model with concave costs and no backlogging. Impose an upper limit of K on the storage capacity; (i.e., $I_n \leq K$ for $n = 2, \ldots, N$). Do parts a to c of Problem 7, where, in part c, the two states for period i now have $I_i = 0$ and $I_i = K$.

9. Reconsider the production control problem with backlogging, and suppose that (5-13) holds for each n and integer x, including $x = -1$. Redefine the active arcs to show that it is optimal to concentrate all production in one period.

10. Consider Program 2 for the case of convex costs. Let Program $2(i, j)$ be the variant of Program 2 in which U_{ij} is increased by 1, with all other data unchanged.

    a. Show that an optimal flow **x** for Program 2 remains optimal for Program $2(i, j)$ if either of the following conditions holds; $x_{ij} < U_{ij}$ or $x_{ij} = U_{ij}$ and the length of the shortest elementary chain from node j to node i the network in Exercise 5 is at least as large as $c_{ij}(x_{ij} + 1) - c_{ij}(x_{ij})$.

[**Hint:** Apply Theorem 5.]

    b. Find an optimal flow **x** for Program $2(i, j)$ when the conditions in part a fail.

[**Hint:** Apply Theorem 5.]

11. Consider Program 2 for the case of convex costs. Call node p *circuit essential* to directed network (S, A) if this network contains at least one simple circuit and if every simple circuit contains node p. (For instance, node 0 is circuit-essential to the directed network in Figure 5-2.) Consider a directed network (N, A) in which node p is circuit-essential.

    a. Fix node $t \neq p$. Define subsets A_+, A_- and A_0 of A so that every elementary chain from p to t has forward arcs in A_+, reverse arcs in A_-, and no arcs in A_0. Show that $A = A_+ \cup A_- \cup A_0$. Show that an optimal flow for Program 2_{pt} (which has b_p increased and b_t decreased by one) differs from an optimal flow for Program 2 by increasing by 1 flows in selected arcs in A_+ and decreasing by 1 selected flows in A_-. Do the same for Program $(2_{pt})_{pt}$.

    b. Fix nodes $s \neq p$ and $t \neq p$. Color nodes s and t red. If an elementary chain exists from s to t that does not include node p, color all intermediary nodes on this elementary chain blue. Then let A_I (I is for indeterminant) be the set of arcs in elementary chains from p to blue nodes that have no colored intermediary nodes. Partition arc set A into disjoint sets A_+, A_-, A_0, and A_I so that every elementary chain from s to t has forward arcs in A_+, reverse arcs in A_-, no arcs in A_0, and arcs whose orientation cannot be foretold in A_I. Relate the optimal flow to Program s_{st} to the optimal flow for Program 2 and to these sets.

Note: Part a of Problem 11 is adapted from Veinott (1973).

6^X

A MARKOV

DECISION MODEL

Titles above in bold type concern advanced
or specialized subjects. These sections can
be omitted without loss of continuity.

UNCERTAINTY

Uncertainty about the future lies at the heart of many decision problems. A retailer stocks products without knowing how many customers will demand each of them. Equipment is purchased without exact knowledge of its maintenance costs or of its useful lifetime. An oil well is drilled without knowing the extent to which it will be successful. An investor does not know which stocks will go up or down. Insurance is bought and sold because of our ignorance of the future.

To say that the future is uncertain is not to say that nothing is known about it. A geologist can estimate the probability that a well will strike oil. An investor can estimate the probability that a venture will succeed. Actuaries compute probabilities relevant to insurance.

A MODEL OF DECISION MAKING
UNDER UNCERTAINTY

When these probabilities can be assessed, rational decision making becomes possible. We now describe a model of decision making in the face of uncertainty. Throughout this chapter, a "period" is an interval of time, and an "epoch" is the beginning of a period. Decision making occurs at finitely many epochs that are numbered 1 through N. Stage n consists of all states that can be observed at epoch n. Generally, transitions occur from epoch n to epoch $n + 1$. The exception is that transitions occur from epoch N to nowhere at all. This model has stages; stage 1 marks the beginning of decision making, and stage N marks the end.

Transition Probabilities

A state in stage n is denoted by a pair (n, i). Associated with each state (n, i) is a nonempty set $D(n, i)$ of decisions. Should state (n, i) be observed, some decision k in $D(n, i)$ must be selected. Provided that $n < N$, this causes transition to some state in stage $n + 1$, the particular state being determined as follows:

$P_{ij}^k(n) = $ the probability that the state observed next is $(n + 1, j)$ if the state observed now is (n, i) and if decision k is selected

Evolution from state to state is governed by transition probabilities, which would also be the case if we were studying a Markov chain. The model we are studying is a blend of Markov chains and decision making that is aptly called a Markov decision model. Although our decision maker faces an uncertain future state, our model presumes that he knows the transition probabilities. The notion that transitions occur from each stage to the next is expressed as follows. For all states (n, i) having $n < N$ and for all decisions k in $D(n, i)$,

$$\sum_j P_{ij}^k(n) = 1$$

where the summation occurs over all j such that $(n + 1, j)$ is a state.[1]

Random Income

No rewards have yet been specified. A reward is earned in each period. Let

$R_i^k(n) = $ the reward earned during period n if state (n, i) is occupied and if decision k is selected

Reward $R_i^k(N)$ is intended to encompass any possible terminal rewards earned at the end of period N.

The description of this model is consistent with the properties of states in sequential decision processes. For instance, the rewards and the transition probabilities are allowed to depend on the state observed and on the decision selected, but not on prior states and not on prior decisions.

Let us review the chain of events. Some state $(1, i)$ is occupied at epoch 1. The decision maker observes this state. She selects a decision k. The (possibly negative) reward $R_i^k(1)$ is earned. Then transition occurs to some state in stage 2. Prior to epoch 2, the state to which transition will occur is unknown and must be regarded as a random variable. At epoch 2, the decision maker observes the new state. She selects a decision, and the chain of events is renewed.

---

[1] In cases where the range of a summation is omitted, the entire range of the variable is indicated.

Imagine that it is now epoch 1. The reward the decision maker earns in period 1 depends on the state the system occupies and on the decision she selects. What reward will she earn at epoch 2? That depends on the state occupied then, which is as yet unknown. The rewards she will earn at epochs 2 through N are as yet unknown and must be regarded as random variables. So, from her current viewpoint, the total income is a random variable, one that depends on the decisions she selects at the various epochs.

Linear Utility Function

This forces the decision maker to choose between random variables. She is often presumed to select the random variable whose expectation is largest. This has the advantage of simplicity; it replaces a random variable by a single figure of merit, its expectation, and it leads to tractable computations.

However, the expectation is by no means the only rational performance criterion. An individual who maximizes expected income is indifferent between a sure thing and a gamble having the same expectation—for instance, between receiving $1000 for certain and an even chance of winning $0 or $2000. An individual who is indifferent between alternatives like these is said to have a *linear utility function*. Individuals (and they are legion) who prefer the sure thing to the gamble with the same expectation are called *risk-averse*.

We presume here that the decision maker has a linear utility function, so she maximizes the expectation of her total income. A business manager often displays this sort of behavior with respect to decisions affecting a small portion of a business. In fact, one of the main reasons for including several different activities in an enterprise is to provide protection against heavy losses in a single activity. Having taken this step, a business manager may be content to maximize the expectation of the income she earns on each activity, as she has "pooled her risk."

To recapitulate, we presume the decision maker observes the state at each epoch, that she know the rewards and transition probabilities associated with each state and decision, and that she maximizes the expectation of total income. In particular, uncertainty is modeled by random phenomena that are drawn from *known* distributions. Although a retailer need not know in advance the number of customers who will demand each product, she is presumed to know the probability that each given number of sales opportunities will materialize.

Policies and Their Return Functions

Moving toward the analysis of this model, we introduce policies explicitly. A *policy* δ is a decision procedure that specifies a decision for each state.[2] Let $\delta(n, i)$ denote the decision that policy δ species for state (n, i). Necessarily,

---

[2] The cognoscenti will recognize that only stationary nonrandomized policies are considered here, the stage being included as a state variable.

$\delta(n, i)$ is an element of $D(n, i)$. All combinations of decisions are possible, and the *policy space* Δ is defined as the set of all such policies. *Using* policy δ means that decision $\delta(m, j)$ is used if state (m, j) is occupied. The *income function* v^δ for policy δ is defined as

> $v^\delta(n, i) =$ the expectation of the total income earned in
> periods n through N if state (n, i) is occupied
> and if policy δ is used

The income function v^δ for any given policy δ can be specified by a simple recursive process. First, when $n = N$, only one decision remains, and

(6-1) $$v^\delta(N, i) = R_i^k(N) \qquad \text{with } k = \delta(N, i)$$

Now suppose that $n < N$. The expectation of a sum is the sum of the expectations; in particular, $v^\delta(n, i)$ is the sum of the reward earned in period n and the expectation of the income earned in periods $n + 1$ through N. And the expectation of the income earned in periods $n + 1$ through N is the sum over j of the probability of a transition to state $(n + 1, j)$ times $v^\delta(n + 1, j)$. Consequently,

(6-2) $$v^\delta(n, i) = R_i^k(n) + \sum_j P_{ij}^k(n) v^\delta(n + 1, j) \qquad \text{with } k = \delta(n, i)$$

In the case of finitely many states, v^δ can be computed by the following procedure. First use (6-1) for each state in stage N, then use (6-2) for each state in stage $N - 1$, then apply (6-2) to stage $N - 2$, and so on. In the case of infinitely many states, this computation would take infinitely long to execute, and the sum in (6-2) might not even be defined. So we proceed on the following assumption.

> *Finiteness Assumption:* The state space and the decision sets have finitely many elements.

The finiteness assumption will be imposed throughout this section. The model studied here is sometimes called the *finite-state, finite-action* case.

Equation (6-2) makes use of the fact that the expectation of the sum is the sum of the expectations. This property of the expectation leads to tractable computation whenever utilities are additive, as in Problem 4.

A Functional Equation

Suppose that transition occurs to state (n, i). As expectations add, the decision maker should maximize the expectation of her income from epoch n on. Let

> $f(n, i) =$ maximum total expected income obtainable
> during periods n through N if state (n, i) is
> occupied

In mathematical terms,[3]

(6-3) $$f(n, i) = \max_{\delta} \{v^\delta(n, i)\}$$

As in Chapter 2, we call policy π *optimal* if v^π equals f, that is, if $v^\pi(n, i) = f(n, i)$ simultaneously for each state (n, i). So an optimal policy must attain a set of maxima simultaneously. We shall soon see that an optimal policy exists and can be found from the solution to the following functional equation.

(6-4) $$g(n, i) = \begin{cases} \max_k \{R_i^k(n) + \sum_j P_{ij}^k(n)g(n + 1, j)\}, & n < N \\ \max_k \{R_i^k(N)\}, & n = N \end{cases}$$

The principle of optimality asserts that $g = f$. To see why, interpret $g(n + 1, j)$ as the best return obtainable when starting in state $(n + 1, j)$. In (6-4), this quantity is multiplied by $P_{ij}^k(n)$, summed over j, and added to $R_i^k(n)$. So the term in braces in (6-4) is the best one can do when starting in state (n, i) and selecting decision k.

Exercise 1. Which version of the principle of optimality was just invoked? Suppose that $R_i^k(n)$ is a random variable rather than a fixed datum. Alter (6-1) to (6-4) accordingly.

Arguments based on the principle of optimality are illuminating, but they are somewhat lacking in rigor.

THEOREM 1. Under the finiteness assumption, (6-4) has exactly one solution, and that solution is f. Moreover, an optimal policy exists, and policy π is optimal if and only if

(6-5) $$g(n, i) = \begin{cases} R_i^k(n) + \sum_j P_{ij}^k(n)g(n + 1, j), & k = \pi(n, i), n < N \\ R_i^k(N), & k = \pi(N, i), n = N \end{cases}$$

Remark: Compare (6-5) with (6-4), and note that π is composed of the decisions that attain the maxima. This theorem is proved by induction on n, downward, in opposition to the direction of the transitions. The proof makes use of the fact that transition probabilities are nonnegative.

Proof. It is first argued that (6-4) has exactly one solution. Note that $g(n, i)$ is defined in terms of $g(n + 1, j)$ for various j. Note also that the sums and maxima in (6-4) occur over finite sets. Now examine (6-4) in decreasing n, starting with $n = N$, to see that $g(n, i)$ is unambiguously defined.

So (6-4) has exactly one solution, g. As there are finitely many states, all maxima in (6-4) are attained, and there exists at least one policy π satisfying (6-5).

So a policy π exists that satisfies (6-5). Consider any policy π that satisfies

---

[3] When the range of a maximand is omitted, its entire range is indicated [i.e., the expression in braces in (6-3) is maximized over all δ in Δ].

(6-5). Compare (6-5) with (6-1) and (6-2) to see that $v^\pi = g$. This and (6-3) assure $f \geq g$.

So any policy π satisfying (6-5) has $v^\pi = g$. It is now argued that $f = g$. Consider any state (N, i) in stage N. Use (6-3), then (6-1), and then (6-4) as follows.

$$(6\text{-}6) \qquad f(N, i) = \max_\delta \{v^\delta(N, i)\} = \max_k \{R_i^k(N)\} = g(N, i)$$

Now adopt the inductive hypothesis that $f(n + 1, j) = g(n + 1, j)$ for each state $(n + 1, j)$ in stage $n + 1$. (This has just been verified for the case $n + 1 = N$.) The right-hand side of (6-2) contains the term $P_{ij}^k(n)v^\delta(n + 1, j)$. The definition of f and the inductive hypothesis assure that $v^\delta(n + 1, j) \leq f(n + 1, j) = g(n + 1, j)$. This inequality is preserved by multiplying by the nonnegative number $P_{ij}^k(n)$ and then summing over j. So

$$(6\text{-}7) \qquad v^\delta(n, i) \leq R_i^k(n) + \sum_j P_{ij}^k(n)g(n + 1, j) \qquad \text{with } k = \delta(n, i)$$

Maximize the right-hand side of (6-7) over k, obtaining $v^\delta(n, i) \leq g(n, i)$. Now maximize over δ, obtaining $f(n, i) \leq g(n, i)$. We have already shown that $g(n, i) = v^\pi(n, i) \leq f(n, i)$, the last from (6-3). Consequently, $f(n, i) = g(n, i)$ for any state (n, i) in stage n. This completes an inductive proof that $f = g$.

So a policy π exists that satisfies (6-5), and any policy π that satisfies (6-5) is optimal. Now consider any policy π that is optimal (i.e., $v^\pi = f$). Note that $f = g$, and substitute g for v^π throughout (6-1) and (6-2) to see that π satisfies (6-5). ∎

Exercise 2. Suppose (6-1) to (6-3) describe the model, but that $P_{ij}^k(n)$ can be negative. Argue that functional equation (6-4) has a unique solution, g. Argue that $g \leq f$.

Efficiency of Recursive Fixing

Let us compare recursive fixing with enumeration. One could use (6-1) and (6-2) to compute v^δ for each δ and then get f from (6-3). This requires evaluation of the right-hand side of (6-1) or (6-2) for every state-policy pair. Recursive fixing requires similar evaluations of the right-hand side of (6-4), but for every state-decision pair. There are many fewer of the latter. Recursive fixing can be regarded as a system for enumerating the policies implicitly, akin to the *implicit enumeration* schemes of integer programming. Incidentally, reaching is not a viable alternative for this model, whose transitions are uncertain, as it is not possible to "reach" from one node to several.

Discounting the Income Stream

The model whose functional equation is (6-4) gives equal weight to dollars earned in period 1 and in period n. However, a dollar earned earlier can be used earlier and is usually preferred. A simple model of the time value of money results from imagining that earnings can be invested to earn interest.

Suppose that the local bank offers depositors a safe investment with interest. Specifically, for each dollar in an account at a given epoch, the bank pays interest of r dollars at the next epoch. The effect of this is that the capital grows by a factor of $(1 + r)$ at each epoch. In particular, a dollar received at a given epoch has the same value of α dollars received one epoch earlier, where

$$\alpha(1 + r) = 1$$

The number α is called the discount factor.

When discussing the time value of money, we assume that reward $R_i^k(n)$ is earned at epoch n rather than at an unspecified time in period n. In the Markov decision model, the income earned at all epochs but the first is likely to be random. Denote the income earned at epoch n by the random variable X_n. Let m be some earlier epoch. Observe that the decision maker is indifferent between receiving X_n at epoch n and receiving $\alpha^{n-m}X_n$ at epoch m, as the latter can be invested for $n - m$ periods at compound interest. For the same reason, the decision maker is indifferent between receiving the income stream $(X_1, \ldots, X_N)$ and receiving the random income

(6-8)
$$\sum_{n=1}^{N} \alpha^{n-1} X_n$$

at epoch 1. The expression given as (6-8) is called the *present value* at epoch 1 of the income stream. When a linear utility function is used, one maximizes the expectation of the foregoing sum.

Exercise 3. Suppose that the decision maker is operating with borrowed money. What should he use as r? What should he maximize?

A *discounted* Markov decision model has r positive, hence α less than 1. The income function $v^\delta(m, i)$ for a discounted model is now defined as the expectation of the present value at epoch m of the income earned from then on if state (m, i) is occupied and if policy δ is used. That is,

(6-9)
$$v^\delta(m, i) = E(\sum_{n=m}^{N} \alpha^{n-m} X_n)$$

where the expectation is conditional upon occupying state (m, i) and upon using policy δ. To save words, we shall refer to $v^\delta(m, i)$ as the *expected discounted income* for starting at state (m, i) and using policy δ.

Exercise 4. Adapt (6-1) to (6-4) to the discounted case.
[Hint: Simply insert α somewhere.]

Of course, the optimal policy varies with the discount factor. On one extreme, when α is zero, only the immediate income is relevant. When α is 1, immediate and future income are weighted equally.

AN ILLUSTRATION FROM INVENTORY CONTROL

An inventory control problem is now formulated as a Markov decision model. Uncertainty enters the model in the form of the sales forecast.

Retailing Skis

Max de Winter sells skis, retail. The wholesaler's truck appears on the first day of each month, and Max tells the driver how many pairs of skis he wants. The driver unloads the skis, accepts Max's check, and drives off. (In the parlance of inventory control, this is the *immediate delivery* case.) Max plans on buying skis on the first day of these six months: September through February. He sells whatever skis are left over on the first of March at a special end-of-the-winter cash sale.

Max's customers demand immediate delivery; if he cannot satisfy them, someone else will. (This is the *lost-sales* case.) Max knows from experience that whatever is left over on March 1 will be sold on that day. Max cannot be certain about the number of customers who will demand skis during any earlier month, and he has decided that it is appropriate to use random variables to describe the demand for skis during the months of September through February. Number these months 1 through 6, and, for $n \leq 6$, define the random variable D_n as follows.

$$D_n = \text{the total demand for skis during month } n$$

$$p_n(j) = \text{the probability that } D_n \text{ equals } j$$

Max is presumed to know the probability distribution of D_n; that is, he is presumed to know the numbers $p_n(j)$ for $j = 0, 1, \ldots$. This suggests that our model requires an unreasonably large amount of data. But if the random variable D_n has the Poisson distribution, Max need only estimate its expectation, as this one parameter determines $p_n(j)$ for each j. Arrivals at service facilities (like sales counters) are normally Poisson; there exists a limit theorem to this effect, and its hypothesis is weak enough to be satisfied in a great many applications. Although this limit theorem has many uses in operations research, it is surprisingly little known, and we devote the final section of this chapter to it.

Our development applies to any distribution of D_n, not just to the Poisson, so we need not invoke the limit theorem here. On the other hand, we must assume that the demands $D_1, \ldots, D_N$ are independent of each other. This assumption leads to an elegant mathematical treatment, but it does not model the situation in which demands are correlated.

Max is presumed to maximize the expectation of the income on this line of merchandise, rather than some risk-averse criteria. However, he figures that for decision-making purposes he is operating with the bank's money. In other words, he wants to discount his income stream in accord with the bank's interest rate on loans. The data of this decision problem are the probabilities $p_n(j)$ and the following:

$W =$ the wholesale price of skis $S < W < R$

$R =$ the retail price of skis

$S =$ the end-of-the-winter sales price of skis

$r =$ the interest rate per month

$\alpha = 1/(1 + r)$

$M =$ the maximum number of pairs of skis that Max can store in his ski department

$N = 6$, the number of epochs in the planning horizon

Assume that the wholesale price is below the retail price but above the end-of-the-winter sale price; that is, W is less than R, but greater than S. The storage capacity of M pairs of skis is included mainly to satisfy the finiteness assumption of the preceding section. Later, we shall ignore M. It is assumed that interest is due the bank on the last day of the month and that Max's customers settle their accounts then.

This model is peculiar in that only one type of ski is considered. Actually, each type and size of ski gives rise to an inventory control model of the sort under consideration.

A Functional Equation for Inventory Control

It is natural for Max to think in terms of maximizing net profit, but it is traditional in inventory theory to minimize net cost. Net cost is the negative of net profit, and minimizing net cost has precisely the same effect as maximizing net profit. To enable you to relate this development to the literature on inventory theory, we shall yield to tradition and minimize net cost. Of course, a profitable line of merchandise has a negative net cost.

Max's decision problem is now formulated as a sequential decision process. Number the months of September through March consecutively as months 1 through 7. What constitutes a state? Consider the situation at the beginning of month n, with $n \leq N$, just before the delivery truck arrives. Max has some number i of pairs of skis on hand and paid for. For current decision-making purposes, prior orders and sales are irrelevant, once i is known. So let state (n, i) depict the situation of having i pairs of skis on hand and paid for at the beginning of month n, just *prior* to arrival of the delivery truck. Then

$f(n, i) =$ the minimum expected discounted cost over the remaining months, given state (n, i)

Max is really interested in $f(1, 0)$, but he has embedded its calculation in a family of optimization problems. A functional equation is now developed for

$f(n, i)$. State $(N + 1, i) = (7, i)$ represents the situation in which i pairs of skis are on hand and paid for on March 1. The bargain hunters will exhaust this stock immediately. The resulting income is Si, and the cost is its negative.

(6-10)
$$f(N + 1, i) = -Si, \qquad i \geq 0$$ ✓ *boundary condition*

For the more elaborate portion of the functional equation, we imagine that Max observes state (n, i) with $n \leq N$ and that he orders $(k - i)$ pairs of skis, which raises the inventory immediately to k. Demand D_n occurs during the month, and it reduces inventory to $(k - D_n)^+$, which is defined as the maximum of $(k - D_n)$ and 0. Consequently, the expectation of the cost of using an optimal policy for months $n + 1$ through $N + 1$ is

$$Ef[n + 1, (k - D_n)^+]$$

where the expectation is taken with respect to the random variable D_n. To justify functional equation (6-11), which follows, we shall account for the payments and receipts occurring during month n and for the time value of money.

(6-11)
$$f(n, i) = \min_{k \mid i \leq k \leq M} \{(k - i)W + g_n(k) + \alpha Ef[n + 1, (k - D_n)^+]\}, \qquad n \leq N$$

(6-12)
$$g_n(k) = \alpha r k W - \alpha R E \min\{D_n, k\}, \qquad n \leq N$$

The net cost incurred during period n equals the ordering cost, which is $(k - i)W$, plus the interest charge on inventory, which is rkW, less the expected sales revenue, which is the expectation of $R\{\min D_n, k\}$. The latter two terms have been aggregated into $g_n(k)$. Payments and receipts occurring at the end of the month have been multiplied by α, which determines the present value of the cost stream at the beginning of month n. So the payments and receipts are properly accounted for in (6-11) and (6-12). Following are explicit formulas for the expectations in (6-11) and (6-12):

(6-13)
$$Ef[n + 1, (k - D_n)^+] = \sum_{j=0}^{k-1} f[n + 1, k - j] p_n(j) + f[n + 1, 0] \sum_{j=k}^{\infty} p_n(j)$$

(6-14)
$$E \min\{D_n, k\} = \sum_{j=0}^{k-1} j p_n(j) + k \sum_{j=k}^{\infty} p_n(j)$$

The rationale just used to obtain (6-11) invoked the principle of optimality. An alternative would be to cast this model as a Markov decision problem and to use the theorem in the preceding section.

Exercise 5. Define $R_j^k(n)$ and $P_{ij}^k(n)$ so as to express the inventory control model in terms of (6-1) to (6-4).

The following exercises ask you to write functional equations for two of this model's many interesting variants.

Exercise 6. Suppose that a setup cost K is incurred if any nonzero order is placed, and suppose there is a 1-month lag in the delivery of orders. Alter the functional equation to accommodate this.

Exercise 7. Suppose that excess demand is backlogged (i.e., held for filling from future inventory), and suppose that customers pay when they order. Alter the functional equation to accommodate this.

[Hint: Allow $i < 0$.]

Include a reasonable model of a final backlog. Does the total of the sales receipts depend on the policy? $\triangle$

In the parlance of inventory theory, the quantity $(k - i)W$ is called the *ordering cost*, and the quantity $g_n(k)$ is, unfortunately, known as the *holding cost*. The so-called "holding cost" includes the sales receipts and, in many cases, depends more strongly on the sales receipts than on the inventory carrying charges. This terminology, which we shall eschew, arose from a situation like that in Exercise 7, where the sales receipts could be ignored.

Recursive fixing consists of evaluating (6-11) for $n = N$ and each i, then for $n = N - 1$ and each i, and so on, ending with $n = 1$. We shall soon see that this procedure entails many superfluous calculations.

A Streamlined Formulation

Functional equation (6-11) represents a Markov decision model having stages. It arises from imagining that the following dialogue occurs at the beginning of each month. The driver appears and asks Max how many pairs of skis he wants. Max tells him. The driver unloads the skis and then disappears for one month. We now imagine a different dialogue that leads to a more efficient formulation.

One-at-a-time scenario:

Driver: "How many do you want?"
Max: either "At least one more," or, "I have enough."

The idea is that this scenario be applied to each state. Doing so alters slightly the meaning of the states. Consider state (n, i). The scenario is initiated with the driver's interrogation. If Max replies, "I have enough," the driver disappears for 1 month. But suppose that Max says, "At least one more." The driver hands him a pair of skis, and transition occurs instantaneously to state $(n, i + 1)$, as inventory is increased by 1. The scenario is then repeated. Ordering six pairs of skis amounts to replying "At least one more" six times. This scenario leads to the functional equation

$$(6\text{-}15) \qquad f(n, i) = \min \begin{cases} W + f(n, i + 1) \\ g_n(i) + \alpha \mathbf{E} f[n + 1, (i - D_n)^+] \end{cases}$$

The top line of (6-15) is composed of the cost W of ordering 1 unit for immediate delivery and the cost $f(n, i + 1)$ of proceeding optimally from the resulting state. The bottom line of (6-15) accounts for ordering nothing this period; to check this, set $k = i$ in (6-11). The top line should be omitted when i equals M (why?), although this has not been shown explicitly.

Functional equation (6-15) cannot represent a model having stages, as some of its transitions are instantaneous and others take 1 month. Nevertheless, this model exhibits a unidirectional sense of motion, as every transition either increases n or maintains n and increases i. Recursive fixing amounts to evaluating (6-15) in a sequence that opposes the direction of motion—namely in decreasing n and, for each n, in decreasing i. Of course, (6-10) still holds for $n = N + 1$ and (6-15) is only used for $n \leq N$.

Is (6-15) more efficient than (6-11)? Equation (6-15) entails two evaluations per state, and one of these is trivial. Equation (6-11) entails one evaluation for every state and for every possible value of k. So (6-15) is faster, and by a wide margin. Of course, the decision maker can employ (6-15) without actually committing himself (and the driver, who might object) to the one-at-a-time scenario. It is only necessary to *think* in these terms.

Problem 16 contains a different model for which the one-at-a-time scenario is useful. The basic idea is to admit *instantaneous transitions*, as well as those taking one epoch. When the idea applies, it leads to a "streamlined" functional equation that is more quickly solved than the original.

Remark: Examine (6-15) to see that the order quantity for state (n, i) is either zero or exactly one more than the order quantity for state $(n, i + 1)$. So (6-15) presumes that at least one optimal policy has that structure. This raises the possibility, which is quashed in Problem 3, that equations (6-11) and (6-15) might have different solutions.

Eliminating Ordering Costs

Functional equation (6-11) contains the term $-iW$ that is independent of k. Factor it out to obtain

(6-16) $f(n, i) = -iW + \min_{k \,|\, i \leq k \leq M} \{kW + g_n(k) + \alpha E f[n + 1, (k - D_n)^+]\}$

This suggests a change of variables, namely

(6-17) $F(n, i) = f(n, i) + iW$

One can substitute (6-17) into both sides of (6-16). We state the result and leave the details to you.

(6-18) $F(n, i) = \min_{k \,|\, i \leq k \leq M} \{G_n(k) + \alpha E F[n + 1, (k - D_n)^+]\}$

(6-19) $G_n(k) = 2kW(1 - \alpha) - \alpha(R - W)E \min \{D_n, k\}$

Exercise 8. Verify (6-18) and (6-19) by substituting (6-17) into (6-16). (Note that $\alpha r = 1 - \alpha$.)

$$kW + g_n(k) - \alpha E(k - D_n)^+ W$$

$$kW + 2\alpha rkW - \alpha R E \min\{D_n, k\} + \alpha \cdot W E \min\{D_n, k\}$$

$$2kW(1 - \alpha) \qquad\qquad -\alpha \cdot W k$$

Recognize (6-18) as a functional equation; it differs from (6-11) in that G_n replaces g_n and in that the ordering cost is eliminated or, more precisely, absorbed into G_n. A description of state (n, i) that is consistent with functional equation (6-18) is provided below.

Exercise 9. Interpret $F(n, i)$ as the cost, including purchase of existing inventory, of starting month n with i pairs of skis on hand and of proceeding optimally from then on. Justify (6-18) and (6-19) directly.

[Hint: In the case $D_n \leq k$, the amount "earned" at the period's end is $RD_n + W(k - D_n)$.]

Functional equation (6-18) applies when $n \leq N$. For the case $n = N + 1$, equations (6-10) and (6-17) combine into

(6-20) $$F(N + 1, i) = i(W - S)$$

Exercise 10. Apply the one-at-a-time scenario to (6-18) to (6-20).

Order-Up-To Levels

The simplest sort of inventory control policy is characterized by *order-up-to levels* $S_1, \ldots, S_N$ in the following way. If the inventory at the start of month n falls below S_n, order enough to restore it to S_n. Otherwise, order nothing.[4] A policy of this sort will be seen to be optimal for the ski-stocking example.

It is now convenient to set $M = \infty$, which allows any nonnegative number of pairs of skis to be ordered. The expression in braces in (6-18) is entirely independent of i. In other words,

(6-21) $$F(n, i) = \min_{k \mid k \geq i} \{L_n(k)\}, \qquad \text{with}$$

(6-22) $$L_n(k) = G_n(k) + \alpha E F[n + 1, (k - D_n)^+]$$

Analysis of $F(n, i)$ turns on the shape of the function $L_n(k)$ of k. This function is shown to be convex in the following theorem, whose hypothesis recaps the assumptions made earlier.

wholesale   *retail*

THEOREM 2. Suppose that $0 \leq \alpha \leq 1$, that $S < W < R$ that $W > 0$, that $M = \infty$, and that ED_n is finite for each n.

For $n = 1, \ldots, N$, the function L_n is convex on $J = \{0, 1, 2, \ldots\}$, and there exists a nonnegative integer S_n such that

(6-23) $$L_n(S_n) = \min_{k \geq 0} \{L_n(k)\}$$

(6-24) $$F(n, i) = \begin{cases} L_n(S_n), & i \leq S_n \\ L_n(i), & i > S_n \end{cases}$$

Proof of the convexity of L_n and of (6-23) are deferred briefly. To see how (6-24) follows from these properties, we examine Figure 6-1. For $k \leq S_n$, the

---

[4] The symbol S_n is traditionally used as the order-up-to level. It is hoped that this will not be confused with stage n.

FIGURE 6-1   Plot of $L_n(k)$ vs. k

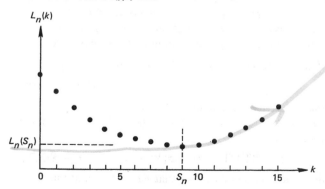

convex function L_n is nonincreasing. For $k \geq S_n$, the convex function L_n is nondecreasing. Consider (6-21) in the light of Figure 6-1 and note that for $i \leq S_n$ the right-hand side of (6-21) is minimized by setting $k = S_n$. Similarly, for $i \geq S_n$, the right-hand side of (6-21) is minimized by setting $k = i$. In short, the optimal ordering rule takes the form of a

Single-critical-number policy: If state (n, i) is observed, set $k = \max\{i, S_n\}$.

The proof of Theorem 2 follows a pattern that is typical of inventory theory. It proceeds by induction, downward on n. It uses Supplement 2. It is intricate enough that it should be overlooked by those readers who are more interested in results than in techniques.

Proof of Theorem 2.   The proof is divided into three parts.

Claim 1.   For $n = 1, \ldots, N$, the function G_n is convex.

Proof.   Equation (6-19) specifies $G_n(k)$ as the sum of the linear term $2kW(1 - \alpha)$ and the term

(6-25) $$-\alpha(R - W)\mathbf{E} \min\{D_n, k\}$$

The hypothesis of Theorem 2 assures that $\alpha(R - W) \geq 0$. Hence, convexity of G_n is equivalent to convexity of the function $\mathbf{E}[-\min\{D_n, k\}]$ of k. Fix D_n, and regard $-\min\{D_n, k\}$ as a function of k. Notice, perhaps by drawing a picture, that this function is convex. Since a convex combination of convex functions is convex, $\mathbf{E}[-\min\{D_n, k\}]$ is a convex function of k. Consequently, G_n is convex, and Claim 1 is proved.

Claim 2. L_N is convex, and $L_N(k) \to \infty$ as $k \to \infty$.

Proof.   Substitute (6-20) into (6-22) to get

(6-26) $$L_N(k) = G_N(k) + \alpha(W - S)\mathbf{E}(k - D_N)^+$$

Claim 1 shows that G_N is convex, and the hypothesis of Theorem 2 assures that $\alpha(W - S) \geq 0$. Hence, the convexity of L_N is assured if we can show that

$E(k - D_N)^+$ is a convex function of k. Fix D_N and notice, perhaps by drawing a picture, that $(k - D_N)^+$ is a convex function of k. Hence, $E(k - D_N)^+$ is convex because it is a convex combination of convex functions. So L_N is convex.

To examine $L_N(k)$ as $k \longrightarrow \infty$, combine (6-19) and (6-26) into

$$(6\text{-}27) \qquad L_N(k) = 2kW(1 - \alpha) - \alpha(R - W)E \min \{D_N, k\} +$$
$$\alpha(W - S)E(k - D_N)^+$$

The hypothesis of Theorem 2 assures that $\alpha(R - W) \geq 0$ and $\alpha(W - S) \geq 0$. Also, $\min \{D_N, k\} \leq D_N$ and $(k - D_N)^+ \geq (k - D_N)$. Combining these inequalities with (6-27) gives

$$(6\text{-}28) \qquad L_N(k) \geq k[2W(1 - \alpha) + \alpha(W - S)] - \alpha(R - S)ED_N$$

The hypothesis of Theorem 2 assures that $ED_N < \infty$ and that the coefficient of k in (6-28) is positive, even when $\alpha = 0$ and when $\alpha = 1$. Hence, $L_N(k) \longrightarrow \infty$ as $k \longrightarrow \infty$. This proves Claim 2.

Claim 3. L_n is convex, and $L_n(k) \longrightarrow \infty$ as $k \longrightarrow \infty$.

Proof. Claim 2 verifies Claim 3 for the case $n = N$. Suppose, inductively, that Claim 3 is true for n. So L_n attains its minimum at some integer S_n that satisfies (6-23), and $F(n, i)$ satisfies (6-24). Hence, $F(n, i)$ is a nondecreasing function of i, and $F(n, i) \longrightarrow \infty$ as $i \longrightarrow \infty$. Now examine

$$(6\text{-}29) \qquad L_{n-1}(k) = G_{n-1}(k) + \alpha EF[n, (k - D_{n-1})^+]$$

Claim 1 shows that G_{n-1} is convex. Also, $\alpha \geq 0$. So convexity of L_{n-1} will be assured if we can show that $EF[n, (k - D_{n-1})^+]$ is a convex function of k.

Since a convex combination of convex functions is convex, it suffices to show that, for each fixed value of D_{n-1}, the function $F[n, (k - D_{n-1})^+]$ is a convex function of k. Fix D_{n-1}, and notice, perhaps by drawing a picture, that $(k - D_{n-1})^+$ is a convex function of k. Since $F[n, \cdot]$ is convex and nondecreasing, Lemma 5 of Supplement 2 shows that, with D_{n-1} fixed, $F[n, (k - D_{n-1})^+]$ is a convex function of k. Hence, L_{n-1} is convex.

To complete an inductive proof of Claim 3, we must show that $L_{n-1}(k) \longrightarrow \infty$ as $k \longrightarrow \infty$. Pick m large enough that $\text{Prob} \{D_{n-1} \leq m\} \geq 0.9$. Then consider $k > m + S_n$. In the case $D_{n-1} \leq m$, we have $(k - D_{n-1})^+ \geq k - m > S_n$, and the fact that $L_n(i)$ is nondecreasing for $i \geq S_n$ assures that $F[n, (k - D_{n-1})^+] \geq L_n(k - m)$. In the case $D_{n-1} > m$, the fact that $F(n, \cdot)$ is nondecreasing assures that $F[n, (k - D_{n-1})^+] \geq F[n, 0]$. Hence,

$$(6\text{-}30) \qquad EF[n, (k - D_{n-1})^+] \geq 0.9L_n(k - m) + 0.1F[n, 0]$$

for $k \geq m + S_n$. Note from (6-19) and the hypothesis of Theorem 2 that

$$(6\text{-}31) \qquad G_{n-1}(k) \geq 2kW(1 - \alpha) - \alpha(R - E)ED_{n-1}$$

Substitute (6-30) and (6-31) into (6-22) to get, for $k > m$,

$$L_{n-1}(k) \geq 2kW(1 - \alpha) - \alpha(R - W)ED_{n-1} + 0.9\alpha L_n(k - m) + 0.1\alpha F(n, 0)$$

Hence, $L_{n-1}(k) \rightarrow \infty$ as $k \rightarrow \infty$, even when α equals 0 and when α equals 1. This proves Claim 3. It also proves Theorem 2, since (6-23) and (6-24) were established in Claim 3. ■

SUMMARY

This chapter introduces decision making in the face of an uncertain future. The finite-state finite-action Markov decision model is analyzed in Theorem 1, and the specific structure of an inventory control model is analyzed in Theorem 2. Both analyses proceed by induction, back from the end of the planning horizon toward the beginning.

The notions of states, stages, embedding, functional equations, and recursive fixing carry over from the deterministic model. Reaching does not. The inventory control model has a formulation with stages and a more efficient formulation that lacks stages.

This chapter's final, starred section contains the surprisingly general circumstances under which the Poisson distribution arises in operations research models.

WHEN TO EXPECT THE POISSON*

It was asserted in "An Illustration from Inventory Control" that the number of customers who arrive at a service facility during a period has, typically, the Poisson distribution. Under restrictive conditions, this distribution is exactly Poisson; under fairly general conditions it is approximately Poisson. Both sets of conditions are described here.

Superposition of Point Processes

These conditions involve the superposition of point processes, a topic best introduced by an illustration. Suppose, for simplicity, that the local market serves only three customers. Consider Figure 6-2. Each "×" on the top line depicts a time at which A. Fishbone arrives at the market. Each "×" on the second line marks a time at which C. Bass arrives. Each of the four lines represents a *point process*, and the fourth is the *superposition* of the top three. The common work "sum" would have done as well as "superposition," as each × on line 4 depicts a point in time at which any customer arrives. The function $N(t)$ equals the number of customers who have arrived by time t. The proprietor is concerned primarily with information concerning $N(t)$, as this affects sales, staffing, stocking, and congestion.

* This section may be omitted without loss of continuity.

FIGURE 6-2. The superposition of point processes

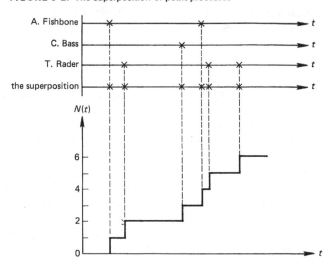

The interarrival times of individual customers are usually random; so $N(t)$ is usually a random variable. The most widely known and least plausible model of this situation is based on the hypothesis that the individual arrival processes satisfy the following equations:[5]

(6-32)
$$\text{Pr \{one ``}\times\text{'' on line } i \text{ between time } t$$
$$\text{and time } t + \Delta t\} = \lambda_i \, \Delta t + o(\Delta t)$$

(6-33)
$$\text{Pr \{more than one ``}\times\text{'' on line } i \text{ between time } t$$
$$\text{and time } t + \Delta t\} = o(\Delta t)$$

Line 4 is the sum of lines 1, 2, and 3. If lines 1 to 3 satisfy (6-32) and (6-33), so does line 4, with $\lambda_4 = \lambda_1 + \lambda_2 + \lambda_3$. Abbreviate λ_4 to λ. By letting Δt approach 0, one obtains a differential equation from (6-32) and (6-33), and a well-trod path leads from this differential equation to the conclusion that

(6-34)
$$\text{Pr } \{N(t) = k\} = e^{-\lambda t}\frac{(\lambda t)^k}{k!}, \qquad k = 0, 1, \ldots$$

Fix t momentarily; a random variable $N(t)$ that satisfies (6-34) is called *Poisson*. When (6-34) holds for each $t \geq 0$, the family of random variables is called a *Poisson process*. The constant λ_i applied to line i is known as its *arrival rate*.

We have concluded that the superposition is a Poisson process, but under an untenable hypothesis. The same arrival rate λ_i was applied to line i at all times—days, nights, weekdays, weekends, Mondays, and Fridays. People are more likely to be shopping at some times than at others. So it might be reasonable to replace λ_i by $\lambda_i(t)$, thereby allowing the arrival rate to vary with time. The

[5] In (6-32), by $f(t) = o(\Delta t)$ we mean $f(t)/\Delta t \longrightarrow 0$ as $\Delta t \longrightarrow 0$.

effect on (6-34) of replacing λ_t by $\lambda_i(t)$ is to replace the product λt with the function $\Lambda(t)$, where

(6-35) $$\Lambda(t) = \sum_{i=1}^{3} \int_{x=0}^{t} \lambda_i(x) \, dx$$

Khinchine (1960) and Çinlar (1975) provide elementary proofs (by changes of variables) that (6-34) remains valid when λt is replaced by $\Lambda(t)$. The linear function λt identifies a *stationary* Poisson process, and a nonlinear function $\Lambda(t)$ identifies a *nonstationary* Poisson process. Of course, these results hold for any number of customers, not just three.

The variable arrival rate $\lambda_i(t)$ allows the arrival probabilities to vary with time, but not with the time of the last arrival. That is, the same probability of arrival applies to customers who just left and to customers who have been gone a week. So the nonstationary Poisson is *memoryless*, a fact that limits its scope.

A Limit Theorem

Ah, but a surprisingly little known limit theorem applies here. The hypothesis of this theorem is expressed (crudely) as the following four conditions.

1. The number n of customers is large.
2. Any single customer accounts for a small fraction of the expected number of arrivals.
3. The customers' interarrival times are independent of each other.
4. Each customer's interarrival times are independent and identically distributed; they also contain some randomness, rather than being fixed.

This theorem's conclusion is that $N(t)$ is approximately a Poisson process. "Approximately" Poisson means that convergence occurs as $n \to \infty$. The rate at which this convergence occurs is proportional to $1/n$, which compares *favorably* with the rate $1/\sqrt{n}$ of convergence in the familiar central limit theorem. This theorem shows that $N(t)$ is approximately a Poisson process, but not necessarily a stationary Poisson process. A precise statement of this theorem and a proof of it can be found in a survey by Çinlar (1972).

Cohorts and the Compound Poisson

Let us check whether or not these four conditions are satisfied at service facilities. Typically, the population n from which customers are drawn runs into the thousands, and even the most frequent user of the facility accounts for a tiny fraction of the arrivals. The interarrival times of individual customers vary, even in cases when they are attempting to follow definite habits. Customers who arrive in different vehicles tend to arrive at different and unrelated times.

But a cohort of customers who arrive in the same vehicle arrive together and violate condition **3**. If this happens frequently, the theorem will not apply—at least not to total arrivals. It still applies to arrivals of cohorts. One need assess the probability distribution of the size of the cohort that arrives together and convolve this distribution with the Poisson distribution. The result is the so-called *compound* Poisson. This situation has an obvious implication for data collection; one should examine the size of the cohorts.

Data and Estimates of the Arrival Rate

The easiest way to take data on an arrival process is to break time into equally spaced epochs and count the number of arrivals in each epoch. Suppose that this has been done and that the sample mean $\hat{m}$ and sample variance $\hat{\sigma}^2$ have been computed. It is not infrequently the case that the sample variance exceeds the sample mean by a factor of 2 or 3. This is sometimes accepted as proof that the arrival process is not Poisson, as the Poisson distribution has its variance equal to its mean.

But a theorem cannot be denied unless its hypothesis can be refuted. If the hypothesis is reasonable, the variance/mean ratio tells the practitioner that he is sampling from a Poisson process that is compound, nonstationary, or both. The data advise him to inquire as to why the variance exceeds the mean. If he can explain this, he can make better use of the data.

Suppose, however, that the practitioner cannot explain why the variance is well in excess of the mean and suppose, in addition, that he can reject the hypothesis of a compound Poisson. (The compound hypothesis can be rejected a priori in cases when it would imply communication between inanimate objects.) The truth as he knows it is that the arrival process is Poisson with an uncertain parameter λ. The uncertainty in λ can be expressed in terms of a probability distribution, and the sample data can be used to assess the parameters of this distribution. The most tractable fit of this type is obtained if one presumes that λ has the gamma distribution, whose density $f(\lambda)$ is expressed in terms· of the parameters r and t by

(6-36) $$f(\lambda) = e^{-\lambda t}(\lambda t)^{r-1} t / \Gamma(r)$$

The results of fitting r and t to the sample mean $\hat{m}$ and sample variance $\hat{\sigma}^2$ are

(6-37) $$t = \frac{1}{(\hat{\sigma}^2/\hat{m}) - 1}, \qquad r = \hat{m}t$$

These readers who are familar with Bayesian statistics will recognize that we are now poised on the brink of a discussion of subjective probability. Instead of plunging into that topic, we refer the interested reader to Problem 20. We do, however, note here that, if r and t are to be updated by Bayes' rule, the dimension of the state space increases by two, and this influences the form of the optimal policy.

BIBLIOGRAPHIC NOTES

The earliest papers on Markov decision processes are cited in Chapter 1. Following Bellman (1957b), Howard's seminal book (1960) sparked intense interest in Markov decision processes, and a vast literature has developed. We cite only the papers that relate directly to this chapter. Work on single-critical-number policies include Bellman, Glicksberg, and Gross (1955), Karlin (1960), and Veinott (1965, 1966b). The notion that the ordering cost can be absorbed into the holding cost may be implicit in Beckmann (1961); it is explicit in Veinott and Wagner (1965). An excellent survey of the early contributions to inventory theory may be found in Scarf (1963). The notion that models can be streamlined by introducing instantaneous transitions occurred to several individuals, including deGhellinck and Eppen (1967) and this writer (1968b).

PROBLEMS

Note: Problems 1 to 5 relate to technical points in the first three sections.

1. Take one side or the other of this proposition: functional equation (6-11) is incorrect because it fails to account for Max's ability to earn interest on the profits he obtains during the early months. Justify your position.
[Hint: Reflect on (6-8) and (6-9).]

2. It was presumed through our discussion of decision making under uncertainty that Δ is the set of all policies δ having $\delta(n, i)$ in $D(n, i)$ for each state (n, i). Does Theorem 1 remain true when Δ is replaced by one of its subsets?

3. Let Δ denote the set of all policies for the ski-stocking example, whose functional equation is (6-11). Let Δ^* be the set of those policies δ in Δ such that, for each state (n, i), the quantity $\delta(n, i)$ equals i or $\delta(n, i + 1)$.

   a. Show that the policy π obtained by solving (6-15) is in Δ^*.

   b. Show that (6-11) and (6-15) have the same solution, f.
[Hint: Induct downward on n and on i.]

   c. Conclude that the ski-stocking example has an optimal policy in Δ^*.

4. An intertemporal utility function $U(X_1, \ldots, X_N)$ is called *additive* if there exist functions $u_1, \ldots, u_N$ such that

$$U(X_1, \ldots, X_N) = u_1(X_1) + \ldots + u_N(X_N)$$

As the expectation of the sum is the sum of the expectations,

$$EU(X_1, \ldots, X_N) = \sum_{n=1}^{N} Eu_n(X_n)$$

Adapt the development in the section on decision making under uncertainty to the case of additive utility. Need $u_n(x)$ be nondecreasing in x?

5. This problem concerns the Markov decision model. Consider this linear program:

$$\text{Minimize } \{\sum_i \sum_n v_i^n\}$$

subject to the constraints

$$v_i^n - \sum_j P_{ij}^k(n)v_j^{n+1} \geq R_i^k(n), \quad \text{all } i, k, n$$

a. Discern an exact relationship between the solution to this program and the optimal return function, f.

[Hint: Is f feasible?]

b. Write the dual of this program. Interpret the optimal dual basis.

6. Bonnie Fields, the debutante, is just 17 and enjoys dating. She figures that she will obtain D_i units of pleasure dating at age i, for $i = 17, \ldots, 30$. She figures that at most one marriage proposal will occur per year, and she estimates $p_i(W)$ and $p_i(F)$ and $p_i(B)$ as, respectively, the probabilities that at age i she will receive a marriage proposal that will turn out Well, Fair, and Bad. She figures that a husband will live until she is 65 and that the three types will provide W, F, and B units of pleasure annually. Show how to calculate her pleasure-maximizing policy.

7. Demand for electricity is increasing by random factors. The demand D_{n+1} during period $n + 1$ is related to D_n by $D_{n+1} = (1 + I_n)D_n$, where I_n is a Poisson random variable with mean 0.1. Initial capacity is zero, and the cost of adding x units of capacity at the start of period n is $c_n(x)$. Excess demand during period n is satisfied by purchase from the regional energy pool at (marginal) unit cost e_n. Excess capacity is sold at that price. The problem is to satisfy demand during periods 1 through N at minimal cost, with D_1 known initially and D_n revealed at the start of period n. Formulate this for solution by dynamic programming.

8. Consider the following variant of the inventory control problem whose functional equation was (6-10) and (6-11). Suppose that the bank pays interest at the rate r_1 per month on deposits and charges interest rate r_2 on loans, where $r_1 < r_2$. Suppose also that at the beginning of the planning horizon, Max has no liquid assets. Alter the functional equation appropriately, still maximizing the present value of his income stream.

9. *Ms. New York* magazine has room for six articles, and its circulation manager, Ann Thrope, wonders what mix of articles maximizes sales. The marketing department classifies articles into K distinct types. The first article of type i generates a Poisson number of sales with mean λ_i. A second article of type i generates a Poisson number of additional sales with mean $\lambda_i/3$, and a third article generates no additional sales. What allocation maximizes sales?

10. Sterling Goldfine thinks he may have discovered the region in which a Spanish galleon sank with a treasure worth $10 million. He figures that the probability is p that he has the right general location. If the treasure is there, he finds it with probability q in one dive. Dives are independent. Each dive costs c thousand dollars. He has enough capital for N dives.

a. Write a functional equation whose solution yields the policy that maximizes his expected wealth.

[Hint: You may wish to compute the conditional probability p_i that the gold is there if there have been i unsuccessful dives.]

b. Investigate the relationship of $c - p_i q 10^4$ to the problem's solution.

11. Foster Sayles, the automobile dealer, is wondering how many cars to order each month from the manufacturer. The wholesale and retail prices are, respectively, W and R per car. The number D_n of customers wanting cars during month n is a Poisson random variable with mean λ_n. The inventory carrying cost is I per car per month. An order placed at the beginning of a month is delivered at the beginning of the next month, and excess demand during the month is lost. The factory wishes to smooth production, so it imposes a special charge of K each month in which Foster changes his order quantity from that in the preceding month. Whatever cars remain at the end of month 12 are sold at an end-of-model-year sale at price W. Show how to compute the number of cars to order each month.

12. Wilde & Katz, the speculators, have taken options on six drilling sites. It costs K dollars to drill at any single site. Site i yields no oil with probability p_i, $3K$ dollars worth of oil with probability q_i and $10K$ dollars worth of oil with probability $1 - p_i - q_i$. They have enough cash to drill at two wells. Any profits from these wells can be reinvested in further drilling. Show how to calculate whether to drill and where to drill.

13. Howie Slams, the tennis pro, has analyzed his service. He has only two serves, the "smash" and the "spin." Let p_1 (and p_2) denote the probability that his smash (respectively, his spin) is in-bounds. Let w_1 (and w_2) be the probability that he wins the point if his smash (respectively, his spin) is in-bounds. Howie wonders whether he should slam on the first and second try.

a. Write the appropriate functional equation and solve it for these data: $p_1 = w_1 = 0.7$ and $p_2 = 0.9$, $w_2 = 0.5$.

b. (*sensitivity analysis*) Over what range of p_1 is the solution in part a optimal?

c. Should his policy depend on the score?

14. In this simplified version of blackjack, aces always count as 1, and cards are dealt from the top of enough shuffled decks that the probability of getting any particular card is essentially independent of the cards that preceded it.

a. Suppose the dealer's first card is up and that it is an i. Let $Q_i(j)$ denote the probability that once during the hand her cards total j. She draws on 16 and sticks on 17. Argue that $Q_i(j) = 0$ for $j < i$, that $Q_i(i) = 1$, that

$$Q_i(j) = \sum_{k=1}^{9} \frac{Q_i(j-k)}{13} + \frac{Q_i(j-10)4}{13}, \qquad i < j \le 16$$

$$Q_i(16+n) = \sum_{k=n}^{9} \frac{Q_i(16+n-k)}{13} + \frac{Q_i(6+n)4}{13}, \qquad n \ge 1$$

   b. Write a functional equation whose solution tells you when to draw and
      when to stick.

[Hint: Ignore all fancy features, such as splitting and five cards under 21.]

   c. Show how to compute the probability of beating the dealer in this version
      of blackjack.

15. Wiley Gaines, the quarterback, needs a touchdown to win. It is first and ten
    on his 20-yard line. There is time for one more touchdown drive. He has K
    plays, numbered $1, \ldots, K$. Play k gains i yards with probability p_i^k, where
    $i = -10, \ldots, 0, 1, \ldots$. Also, $\sum_i p_i^k$ is less than 1 by the probability q^k
    that play k results in loss of the ball through a fumble or interception. Show
    how to compute his optimal policy.

16. Consider the variant of the ski-stocking example in which Max de Winter
    incurs setup cost K if he places any nonzero order (e.g., the cost of ordering
    three pairs of skis is now $K + 3W$).

    a. Write a functional equation whose solution specifies Max's optimal
       ordering strategy.

       The four figures shown below reflect ways in which this functional
    equation might be streamlined. The integer n has been suppressed every-
    where [e.g., stock level $(n, 0)$ becomes 0]. The thick-stemmed arrow reflects
    no (further) order, in which case transitions occur from stage n to stage
    $n + 1$ in the same manner as in (6-15). The arrow connecting two nodes
    reflects an instantaneous transition whose cost is adjacent to it.

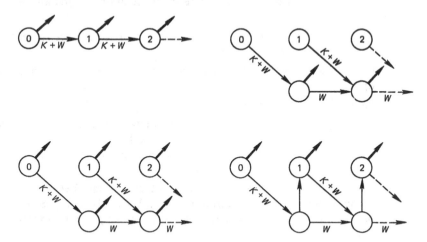

    b. Which of the four figures is incorrect? Why?

    c. Write functional equations for the correct figures.

    d. If multiple figures are correct, which functional equation is the most
       efficient. Why?

17. What was the name of Max de Winter's first wife?

> *Note:* Problems 18 to 21 relate to the section on the superposition of point processes.

18. Let N be a Poisson random variable with parameter λ; that is, $\Pr\{N = k\} = e^{-\lambda}(\lambda)^k/k!$ for $k = 0, 1, \ldots$. Show that $EN = \lambda$, that $EN(N - 1) = \lambda^2$, that $EN^2 = \lambda^2 + \lambda$, and that Var $(N) = \lambda$.

19. (*Variance/mean ratio*) Let N_1 and N_2 be Poisson random variables with parameters λ_1 and λ_2, respectively. Let Z be a random variable with $p = \Pr\{Z = 1\}$ and $1 - p = \Pr\{Z = 2\}$.

    a. Compute $E(N_Z)$ and Var (N_Z).

[Hint: Condition on Z, and use the results of Problem 18.]

    b. Suppose that $\lambda_1 \neq \lambda_2$. For what values of p are the variance-to-mean ratios greater than one?

    c. Would you expect a situation related to that in part b to arise when data are taken from a nonstationary Poisson process?

20. The gamma function $\Gamma(r)$, which appeared in the denominator in (6-36), is defined by

$$\Gamma(r) = \int_0^\infty e^{-x} x^{r-1} \, dx$$

    a. Show that $\Gamma(r + 1) = r\Gamma(r)$, that $\Gamma(1) = 1$, and for r a positive integer, that $\Gamma(r + 1) = r!$.

[Hint: Integrate by parts.]

    b. Consider $f(\lambda)$ as defined by (6-36). Show that

$$1 = \int_0^\infty f(\lambda) \, d\lambda,$$

$$\frac{r}{t} = \int_0^\infty \lambda f(\lambda) \, d\lambda, \qquad \frac{r(r+1)}{t^2} = \int_0^\infty \lambda^2 f(\lambda) \, d\lambda$$

    c. Suppose that N is a Poisson random variable whose parameter λ has density $f(\lambda)$ given by (6-36). Use the conditional expectation formula, which is $E\Phi(N) = \int_0^\infty E\{\Phi(N)\,|\,\lambda\} f(\lambda) \, d\lambda$, to show that

$$EN = \frac{r}{t}, \qquad \text{Var } (N) = \frac{r}{t}\frac{1+t}{t}$$

[Hint: You may find part b and Problem 18 helpful.]

    d. Justify (6-37).

21. Suppose that N is a Poisson random variable whose parameter λ is uncertain. Postulate that λ has the gamma distribution, and that a manager is asked to provide information that will be used to estimate the distribution of λ.

    The manager feels that $\Pr\{N \geq 9\} = 0.85$ and that $\Pr\{N \leq 11\} = 0.85$. Use this information to estimate the mean and standard deviation of the density $f(\lambda)$, and determine r and t.

[Hint: Use the results of Problem 20.]

7

INVENTORY CONTROL:

(s, S)-POLICIES

INTRODUCTION
AN INVENTORY CONTROL MODEL
ALTERNATIVE CONDITIONS FOR (s, S)-POLICIES
VARIATIONS ON A THEME
SUMMARY
PROOFS OF OPTIMALITY OF (s, S)-POLICIES √
BIBLIOGRAPHIC NOTES
PROBLEMS

Titles above in bold type concern advanced
or specialized subjects. These subjects can
be omitted without loss of continuity.

INTRODUCTION

This chapter concerns a model of inventory control with uncertain demand and with a fixed charge for placing an order. In each of several periods, the inventory control manager observes the inventory and decides how many units, if any, to order. The ordering cost includes a cost per unit ordered and a fixed charge or *setup cost* for placing any order. Included in the setup cost are the expenses of writing the order, of receiving the merchandise, of writing the check that pays the supplier's bill, and of putting the items received into inventory. To a first approximation, these expenses are independent of the size of the order.

The cost-minimizing ordering rule for this type of model is often characterized by two numbers per period, as follows. Each period n has an *order-up-to-quantity* S_n and a *reorder point* s_n, where $s_n \leq S_n$. If the inventory x at the start of period n is at least s_n, no order is placed. If the inventory x at the start of period n is below s_n, exactly $(S_n - x)$ units are ordered. This two-parameter ordering rule is called an *(s, S)-policy*. When $s_n = S_n$ for each n, the (s, S)-policy reduces to the *single-critical-number* policy introduced in Chapter 6.

An (s, S)-policy has the property that any nonzero order placed in period n must be for more than $(S_n - s_n)$ units. Thus, the minimum order size can be set large enough that the benefit of the added inventory offsets the setup cost. A single-critical-number policy lacks this feature; it orders only one item when the inventory x equals $S_n - 1$, and the benefit of this item may not offset the setup cost.

After presenting two sets of conditions under which (s, S)-policies are optimal, we show how these conditions adapt to different sorts of inventory control situations. Proofs of optimality are placed in the chapter's final section.

AN INVENTORY CONTROL MODEL

A specific inventory control model is now described. The item whose inventory is being controlled is now assumed to be *indivisible* (i.e., it occurs in integer quantities, not in fractions). A *fixed planning horizon* is now assumed (i.e., ordering opportunities occur at the beginning of periods 1 through N). *Instantaneous delivery* of orders is now assumed (i.e., items ordered from the supplier at the start of period n become available instantaneously). *Deterioration of stock* is excluded (i.e., pilferage, spoilage, and balking are assumed not to occur). These assumptions are relaxed subsequently.

The symbols x and y are used consistently throughout this section to describe the controls in this model. Throughout, x denotes the inventory on hand at the start of a period, just prior to deciding whether to place an order. Also, y denotes the inventory on hand at the start of a period, just after deciding whether to place an order. Since delivery is instantaneous, the number of units ordered is $(y - x)$. Negative orders (shipbacks) are not allowed, so $y \geq x$. Since stock is indivisible, x and y are integers. In one version of this model, the excess of demand over stock is held as a backlog to be filled by future orders. In this version, the "inventory" level can take negative values, and a negative value of x indicates that the item is out of stock and that $-x$ units of demand await future orders.

The data of this inventory control model are the constants and functions displayed and described below. When x and y appear as arguments, they hold the meanings described above.

$N =$ the number of periods in the planning horizon. (Ordering opportunities occur at the start of periods 1 through N.)

$K_n =$ the setup cost for period n. (Cost K_n is nonnegative, and this cost is incurred if a nonzero order is placed at the start of period n.)

$W_n =$ the (wholesale) unit ordering cost in period n. (The cost of ordering $z > 0$ units at the start of period n is $K_n + zW_n$.)

$R_n =$ the (retail) unit sales price during period n. (Customers are assumed to pay at the end of the month in which they place their orders *even* if their orders are backlogged.)

$D_n =$ the demand for the item during month n. (Demand D_n is a random variable that takes nonnegative integer values, and the demands in different periods are assumed to be independent random variables.)

$\alpha =$ the one-period discount factor

$h_n(y) =$ the expectation of the inventory carrying cost during period n, given as a function of y. (The actual carrying cost may depend on the random variable D_n, but its expectation with respect to D_n does not. When y is negative, $h_n(y)$ includes the cost of keeping track of the backorders and the monetary equivalent of the loss of good will attendant this backlog.)

$I_n(y, D_n) =$ the inventory on hand at the start of period $n + 1$, given as a function of y and of D_n.

$e(x^+) =$ the salvage value of having $x^+ = \max \{x, 0\}$ units of inventory on hand at the end of period N.

Two differences between this model and the one in Chapter 6 are represented by the functions I_n and e. Including I_n in the specification of this model lets us treat the lost sales, backlogging, and other cases coherently. *Backlogging* means that $I_n(y, D_n) = (y - D_n)$ because the excess, if any, of demand D_n over stock y is held as a backlog against future orders. *Lost sales* means $I_n(y, D_n) = (y - D_n)^+$ because the excess, if any, of demand over stock is lost.

Including e in the specification of the model, let us examine the sorts of terminal conditions that lead to the optimality of (s, S)-policies. If backlogging is allowed, the inventory x at the end of period N can be negative. This depicts a final backlog of $-x$ units for which payment *has* been received. Our model includes a final order that satisfies any excess demand that remains at the end of the planning horizon. Specifically, when the final inventory x is negative, the inventory control manager is compelled to order $-x$ units at cost $K_{N+1} + (-x)W_{N+1}$. So K_{N+1} and W_{N+1} are the setup and unit costs for the final order.

To describe the ordering cost mathematically, we employ the *Heavyside* function $H(\cdot)$, where

$$H(x) = \begin{cases} 0 & \text{for } z \le 0 \\ 1 & \text{for } z > 0 \end{cases}$$

Hence, the cost of ordering z units from the supplier at the start of period n equals $K_n H(z) + zW_n$.

Let $c_n(x, y)$ denote the present value at the start of period n of the expected cost incurred during the period, given in terms of the period's initial inventory x and after-ordering inventory y. An expression for $c_n(x, y)$ is written below and interpreted subsequently.

(7-1) $\quad c_n(x, y) = K_n H(y - x) + (y - x)W_n + h_n(y) - [y - \mathbf{E}I_n(y, D_n)]\alpha R_n$

The elements of (7-1) are described in three parts. The term $K_n H(y - x) + (y - x)W_n$ is the cost of increasing inventory from x to y by ordering $y - x$ units for immediate delivery. The term $h_n(y)$ in (7-1) is the expectation of the cost of carrying for one period the y units of inventory that are on hand at the start of the period. Deterioration of stock is excluded, and the number of units sold during period n equals $[y - I_n(y, D_n)]$ because sales are the excess of y over the inventory $I_n(y, D_n)$ remaining at the start of the next period. The unit sales price is R_n. Customers pay at the end of the period in which they order, and $[y - EI_n(y, D_n)]\alpha R_n$ is the present value at the start of the period of the expectation of the revenue from customers' orders. This is converted into a cost in (7-1) by multiplying it by -1.

   In a dynamic programming formulation of the inventory control problem, state (n, x) denotes the situation of having x units of inventory on hand and paid for at the start of month n, just before deciding whether (and how much) to order. Let X denote the set of attainable inventory levels. Until now X has been the set of nonnegative integers. Here, X is allowed to be any closed subset of the real numbers.

   Interpret $f(n, x)$ as the present value at the start of period n of the cost incurred from then to the end of the planning horizon if state (n, x) is observed and if an optimal policy is followed. This leads to the following functional equation, which is interpreted subsequently.

(7-2) $$f(n, x) = \operatorname*{infimum}_{y|y \geq x} \{c_n(x, y) + \alpha E f[n + 1, I_n(y, D_n)]\}$$

The term $c_n(x, y)$ in (7-2) is the present value of the cost incurred during period n, given in terms of x and y. The quantity $f[n + 1, I_n(y, D_n)]$ is, by definition, the present value at the start of period $n + 1$ of the cost incurred from then to the end of the planning horizon if the inventory at the start of period $n + 1$ equals $I_n(y, D_n)$ and if an optimal policy is employed from then on. Hence, the term $\alpha E f[n + 1, I_n(y, D_n)]$ represents the present value at the start of period n of the costs incurred during the subsequent periods if the inventory at period n is raised to y units and if an optimal policy is employed in all subsequent periods. Consequently, the term in brackets in (7-2) is the discounted cost of raising inventory to y and using an optimal policy thereafter.

   At this point in the development, functional equation (7-2) is supported solely by the principle of optimality (i.e., by our tacit assumption that an optimal policy exists). For instance, Theorem 1 of Chapter 6 fails to apply because this model has infinitely many states and infinitely many decisions. Moreover, the situation is rich enough to admit of various pathologies. For instance, the expectations in (7-2) need not exist, and the infimum need not be finite. Conditions that preclude these pathologies will soon be presented.

   Functional equation (7-2) is incomplete in the case $n = N$ because $f[N + 1, \cdot]$ appears on its right-hand side. Since the inventory left over at the end of the planning horizon is $I_N(y, D_N)$, the functional equation can be completed by adding the terminal condition

(7-3) $$f(N + 1, x) = K_{N+1} H(-x) + (-x)^+ W_{N+1} - e(x^+)$$

The first two terms of (7-3) account for the cost of disposing of excess demand $(-x)^+$ by a special end-of-planning-horizon order. The final term in (7-3) is the salvage value $e(x^+)$, converted to a cost.

A Change of Variables

The term $-xW_n$ on the right-hand side of (7-1) is independent of y and can be factored out of (7-2). This motivates the change of variables

$$(7\text{-}4) \qquad F(n, x) = f(n, x) + xW_n$$

The result of substituting (7-4) into both sides of (7-2) is presented in the three equations that follow, with the algebraic detail left for you to check.

$$F(n, x) = \operatorname*{infimum}_{y \mid y \geq x} \{K_n H(y - x) + G_n(y) + \alpha E F[n + 1, I_n(y, D_n)]\}, \qquad n \leq N$$
$$(7\text{-}5)$$

$$(7\text{-}6) \qquad F(N + 1, x) = K_{N+1} H(-x) + x^+ W_{N+1} - e(x^+)$$

$$(7\text{-}7) \qquad G_n(y) = h_n(y) + (W_n - \alpha R_n)y + \alpha(R_n - W_{n+1}) E I_n(y, D_n)$$

Recognize (7-5) and (7-6) as a functional equation. The quantity $G_n(y)$ is called the *operating cost*; it accounts for all units of cash flow during period n except for the setup cost. The first term in $G_n(y)$ is the inventory carrying cost. To interpret the remaining terms, we imagine that the starting inventory y is purchased from the supplier at unit cost W_n, that ending inventory $I(y, D_n)$ is returned to the supplier at unit price W_{n+1}, and that sales of $y - I(y, D_n)$ units occur at unit price R_n. To obtain (7-6), manipulate these quantities by taking expectations with respect to D_n, multiplying end-of-month transactions by α, and converting revenues into costs by multiplying by -1.

Exercise 1. Verify (7-5) to (7-7) by substituting into (7-1) to (7-3). In addition, interpret state (n, x) in a manner consistent with (7-5).

Having absorbed the linear portion $(y - x)W_n$ of the ordering cost into the operating cost, we separate (7-5) into the pair of equations

$$(7\text{-}8) \qquad F(n, x) = \operatorname*{infimum}_{y \mid y \geq x} \{K_n H(y - x) + L_n(y)\}$$

$$(7\text{-}9) \qquad L_n(y) = G_n(y) + \alpha E F[n + 1, I_n(y, D_n)]$$

Up to this point, the development has followed the same pattern as in Chapter 6, where the convexity of L_n followed (in Theorem 2) from the convexity of G_n. However, when the setup costs are positive, convexity of G_n does not assure convexity of L_n. In fact, convexity of G_n does not preclude the decidedly nonconvex shape of L_n depicted in Figure 7-1. The function depicted in Figure 7-1 has two local minima. The point identified in Figure 7-1 as S_n is the point y at which the function $L_n(y)$ attains its global minimum. The point identified as s_n in Figure 7-1 is the smallest value of y for which $L_n(y) \leq L_n(S_n) + K_n$. Since K_n is nonnegative, one has $s_n \leq S_n$. So S_n and s_n satisfy

$$(7\text{-}10) \qquad L_n(S_n) = \inf \{L_n(y) \mid y \in X\}$$

$$(7\text{-}11) \qquad s_n = \inf \{y \in X \mid L_n(y) \leq L_n(S_n) + K_n\}$$

FIGURE 7-1. The function $L_n(y)$ for $y \in X = \{0, 1, \ldots\}$

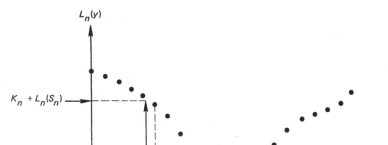

It is now argued that an (s, S)-policy is optimal for the strangely shaped function in Figure 7-1. Consider state (n, x) with $x \geq s_n$; note from (7-8) and Figure 7-1 that it is optimal to set $y = x$ because the setup cost cannot be recouped by increasing y. Now consider state (n, x) with $x < s_n$; note from (7-8) and Figure 7-1 that it is optimal to set $y = S_n$ because the setup cost can be recouped.

Exercise 2. Plot the function $F(n, x)$ of x on Figure 7-1. Is $F(n, \cdot)$ monotone?

Sufficient Conditions

Theorem 1 below gives conditions under which an (s, S)-policy is optimal. This theorem is stated in a way that allows the set X of attainable inventory levels to be a closed subset of the real numbers, rather than being limited to the integers. In the most representative situations, X takes one of the following forms: the integers, the nonnegative integers, the real numbers, and the nonnegative real numbers. Theorem 1 encompasses the case of *divisible* merchandise in which the inventory level can take integer, fractional, and irrational values.

The theorem's hypothesis is lengthy; its proof is intricate. The hypothesis is interpreted immediately after the theorem's statement, and the proof is deferred to a later section.

THEOREM 1. Suppose that X is a closed subset of the reals, and that the following five conditions hold.

(a) (*Convex operating cost*). For $n = 1, \ldots, N$, the function $G_n(\cdot)$ is continuous and convex on X.

(b) (*Nondecreasing present value of setup cost*). For $n = 1, \ldots, N$, $K_n \geq \alpha K_{n+1}$.

(c) (*Convex nondecreasing carry-forward*). For $n = 1, \ldots, N$ and for each value of D_n, the function $I_n(\cdot, D_n)$ is continuous, convex, and nondecreasing on X.

(d) (*Convex salvage cost*). The function $x^+ W_{N+1} - e(x^+)$ is continuous, convex, and nondecreasing in x.

(e) (*Technical*). All expectations are finite. Also, for $n = 1, \ldots, N$.

(7-12)
$$\operatorname*{infimum}_{x \in X} \{L_n(x)\} < \operatorname*{infimum}_{x \in X \,||\, x| \to \infty} \{L_n(x)\}$$

Then, for $n = 1, \ldots, N$, there exists S_n in X that satisfies (7-10), and there exists s_n in X that satisfies (7-11), possibly with $s_n = -\infty$. Moreover,

(7-13)
$$F(n, x) = \begin{cases} K_n + L_n(S_n) & \text{if } x < s_n \\ L_n(x) & \text{if } x \geq s_n \end{cases}$$

Exegesis

The hypothesis of Theorem 1 is now interpreted in the context of indivisible merchandise (i.e., X contains only integers). Every function is continuous on the integers, so the continuity conditions in the hypothesis of this theorem are trivially satisfied. It is now argued that conditions (a) to (e) are satisfied in the simplest, most representative forms of this inventory control model.

Condition (a) is expressed in terms of the operating cost G_n, which is a composite of more primitive quantities. It is broken down in

Exercise 3. Show that (a) is a consequence of (b) to (g), where (f) and (g) are

(f) $h_n(\cdot)$ is convex for $n = 1, \ldots, N$
(g) $R_n \geq W_{n+1}$ for $n = 1, \ldots, N$

Condition (f) holds in the simple model $h_n(x) = x^+ C_n + (-x)^+ D_n$, where the nonnegative constants C_n and D_n are, respectively, the unit holding and shortage costs. Condition (g) holds when the retail sales price R_n is at least as large as the replacement cost W_{n+1}. This condition is natural; if it were violated, the inventory control manager would be motivated to withhold inventory from the market because the sales price is below the replacement cost. Hence, condition (a) flows naturally from the others.

Condition (b) holds when the interest rate is nonnegative (i.e., $\alpha \leq 1$) and the setup cost is independent of the period (i.e., $K_n \equiv K$).

Condition (c) holds in the backlogging case $I_n(y, D_n) = (y - D_n)$ and in the lost-sales case $I_n(y, D_n) = (y - D_n)^+$. In a subsequent section, we shall see that (c) also holds for certain models of balking, spoilage, and deterioration of stock.

In the case of indivisible merchandise, condition (d) can be written in the equivalent form

$$W_{N+1} \geq e(1) - e(0)$$
$$e(x + 1) - e(x) \leq e(x) - e(x - 1), \qquad x = 1, 2, \ldots$$

Hence, (d) presumes that the salvage value of the $(x + 1)$st item is no more than the salvage value of the xth item and that the salvage value of each item is below the wholesale cost. In short, the salvage market is assumed to saturate, and it is assumed to be unprofitable to buy from the wholesaler for the purpose of selling in the salvage market. Similarly, the inequality $R_N \geq W_{N+1}$ in (g) guarantees that it is not profitable to withhold items from customers in order to sell them in the salvage market.

In the lost-sales case, W_{N+1} lacks a natural meaning. We are free to take $W_{N+1} = R_N$, which preserves (g) and reduces part of (d) to the inequality $R_N \geq e(1) - e(0)$. Consequently, in the lost-sales case, (d) can be replaced by the assumption that the salvage-value function e is concave and that the salvage value of 1 unit is no greater than the retail sales price R_N in the final epoch. Hence, it remains unprofitable to withhold items from customers during the final period in order to sell them in the salvage market.

Part of condition (e) is that random variables have finite expectations, which occurs in practical situations. The remainder of condition (e) is (12), which assumes that it is not desirable to allow the inventory level to approach $+\infty$ or $-\infty$. Neither part of (e) would trouble a practitioner.

In short, the hypothesis of Theorem 1 describes a natural situation, one that is likely to arise in practice.

ALTERNATIVE CONDITIONS FOR (s, S)-POLICIES

Other sets of conditions under which (s, S)-policies are optimal are given in this section.

Let L map a subset X of the real numbers into the real numbers. The function L is called *quasi-convex* on X if there do not exist elements $a < b < c$ of X such that $L(a) < L(b) > L(c)$. A quasi-convex function cannot increase and then decrease. A convex function is quasi-convex. Three examples of quasi-convex functions are given in Figure 7-2.

The function L is called *lower semicontinuous* on a closed set X if, for each real number y, the set $\{x \in X \mid L(x) \leq y\}$ is closed. The Heavyside function

FIGURE 7-2. Three quasi-convex functions on $X = \{x \mid x \geq 0\}$

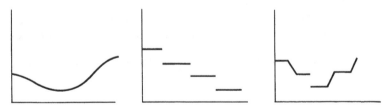

$H(\cdot)$ is lower semicontinuous. Indeed, our need for lower semicontinuity stems from the fact that the Heavyside function, which appears in (7-1), is lower semicontinuous, not continuous. A lower semicontinuous function attains its infimum on any closed bounded subset of X. Two examples of lower semicontinuous functions are given in Figure 7-3.

FIGURE 7-3. Two lower semicontinuous functions on $X = \{x \mid x \geq 0\}$

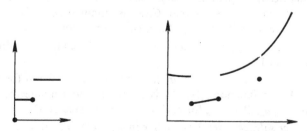

The function L is called *K-convex* on X if all elements $a < b < c$ of X have

(7-14) $$L(c) + K \geq L(b) + (c - b)\left(\frac{L(b) - L(a)}{b - a}\right)$$

Figure 7-4 illustrates K-convexity; it shows the straight line passing through points $[a, L(a)]$ and $[b, L(b)]$ in the plane, and (7-14) requires the height of this line above c to be no more than $L(c) + K$.

Exercise 4. Divide (7-14) by $(c - b)/(c - a)$ to check that 0-convexity is equivalent to convexity.

FIGURE 7-4. K-convexity

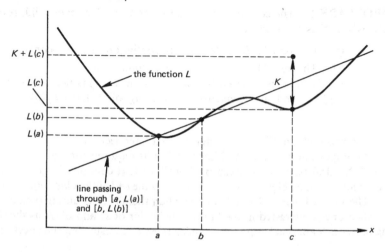

Convexity plays a central role in the hypothesis of Theorem 1. In the following theorem, convexity is relaxed to quasi-convexity, but side conditions are needed.

THEOREM 2.   The conclusions of Theorem 1 remain valid under the following hypothesis.

(a) (*Quasi-convex operating cost*). There exist elements $a_1 \leq \ldots \leq a_N$ of the closed set X such that $G_n(x)$ is nonincreasing for $x \leq a_n$ and nondecreasing for $x \geq a_n$. Also, G_n is continuous on X.

(b) (*Nondecreasing present value of setup cost*). Same as condition (b) of Theorem 1.

(c) (*Nondecreasing carry-forward*). For $n = 1, \ldots, N$ and for each value of D_n, the function $I_n(\cdot, D_n)$ is continuous and nondecreasing on X. Also, $I_n(a_n, D_n) \leq a_n$ for each n and each value of D_n.

(d) (*Linear salvage cost*). For each x in X, $e(x^+) = x^+ W_{N+1}$.

(e) (*Technical*). Same as condition (e) of Theorem 1.

The proof of Theorem 2 is deferred to a later section. An advantage of Theorem 2 is that it allows the operating cost G_n to be quasi-convex, but not convex. It can, however, be hard to demonstrate that a nonconvex function is quasi-convex. An important advantage of Theorem 2 over Theorem 1 is that the carry-forward quantity $I_n(\cdot, D_n)$ is only required to be nondecreasing, not convex. This will be seen to encompass a wider range of models of balking and of deterioration of stock. One limitation of Theorem 2 is its restrictive assumption on the salvage function $e(x)$, which must be linear, with slope W_{N+1}.

The salvage-function assumptions of Theorems 1 and 2 can be removed when one can verify the hypothesis of

COROLLARY 1.   The conclusions of Theorems 1 and 2 remain valid, respectively, when (d) is replaced by

(d)′ (for Theorem 1). L_N is lower semicontinuous and K_N-convex on X.

(d)″ (for Theorem 2). L_N is lower semicontinuous on X, and the condition on G_N in (a) is replaced by the following: $L_N(x)$ is nonincreasing for $x \leq a_N$, and any elements $c > b \geq a_N$ of X have $L_N(b) \leq K_N + L_N(c)$.

The proof of Corollary 1 is deferred to a later section.

An important limitation to Theorem 2 is its requirement that $a_1 \leq \ldots \leq a_N$, and this drawback is preserved by Corollary 1. It connotes larger values of S_n as n increases (i.e., higher stock levels toward the *end* of the planning horizon).

Theorem 2 holds a clear advantage when there is discounting (i.e., $\alpha < 1$) and when one is interested in the limiting behavior of s_1 and of S_1 as the planning horizon N approaches infinity. Under these circumstances, the effect of the end of the planning horizon is attenuated by the factor α^N. Any convenient model of the end of the planning horizon will do, provided that α^N attenuates its cost to zero as N approaches infinity.

Recorded below are general conditions under which single-critical-number policies are optimal.

COROLLARY 2. Augment the hypotheses of Theorems 1 and 2, respectively, by the condition $K_n = 0$. Then an optimal (s, S)-policy exists for which $s_n = S_n$.

Exercise 5. Show that Corollary 2 is a consequence of Theorems 1 and 2. [Hint: Take as S_n the smallest minimizer of L_n.]

VARIATIONS ON A THEME

In this section the inventory control model is adapted to accommodate the following features: payment upon delivery, lag in delivery, balking, and deterioration of stock. Each feature is described as a variation on the inventory control model introduced at the beginning of the chapter. Each of the variants is cast in a form that lets us use Theorems 1 and 2 to demonstrate the optimality of an (s, S)-policy. Hence, this section broadens the class of models having optimal (s, S)-policies. It also broadens the class of models having optimal single-critical-number policies when its results are specialized to the case in which all setup costs are zero.

Backlogging with Payment Upon Delivery

Consider the variant of the inventory control model in which excess demand is backlogged, customers pay at the end of the periods in which they receive their merchandise, and customers pay the price in effect when their orders are filled. For simplicity, suppose that merchandise is indivisible (i.e., $X = \{0, \pm 1, \pm 2, \ldots\}$).

In this case, the inventory control manager faces a new and simpler situation at the end of the planning horizon. If the final inventory x is positive, he can sell it for salvage at price $e(x)$, as before. If the final inventory x is negative, it represents a backlog of $-x$ orders for which he has not accepted payment. Having accepted no money, he has less need to fill the order. Aiming for the simplest model of payment upon delivery, we eliminate the special order at the end of the planning horizon. That is, we simplify (7-3) to

(7-15) $$f(N + 1, x) = -e(x^+)$$

Now, consider the effect of deferred sales receipts. When $D_n \leq y$, the sales receipts during period n remain equal to $D_n R_n$. When $D_n > y$, the sales receipts are deferred until the first period in which merchandise becomes available. The number of deferred sales is $(D_n - y)^+$, and the present value at the start of period n of deferring one sale to the next period is $\alpha(R_n - \alpha R_{n+1})$. Consequently, the expected present value at the start of period n of deferring sales receipts to the next period is given by the expression

(7-16) $b_n(y) = \alpha(R_n - \alpha R_{n+1})E(D_n - y)^+, \qquad n = 1, \ldots, N$

When $n = N$, expression (7-16) contains the term R_{N+1} that has not appeared previously. This model excludes sales at the end of the planning horizon; hence, $R_{N+1} = 0$. The best way to convince oneself that (7-16) accounts correctly for deferred sales receipts may be to work an example, such as the following.

Exercise 6. Show that adding $b_n(y)$ to $G_n(y)$ accounts correctly for sales revenue in this instance: $y_1 = 10$, $D_1 = 13$, $y_2 = -3$, $D_2 = 2$, $y_3 = 15$, and $D_3 = 1$.

Since the final order has been excluded, it may seem natural to set $W_{N+1} = 0$. We shall, however, find another role for the constant W_{N+1}. Equation (7-15) accounts correctly for the end of the planning horizon no matter what value is taken for W_{N+1}. This constant now makes its first appearance in the change of variables defining $F(N + 1, x)$. It reappears in the operating cost $G_N(y)$ and in condition (d) of Theorems 1 and 2. To preserve condition (d), set $W_{N+1} = e(1) - e(0)$ (in this case of indivisible merchandise).

The adaptation for payment upon delivery is now completely specified. It is recapitulated below.

> *Adaptation for backlogging with payment upon delivery:* Add the term $b_n(y)$ to the inventory carrying cost $h_n(y)$, replace (7-3) by (7-15), and set $W_{N+1} = e(1) - e(0)$.

We now consider whether this adaptation preserves the hypothesis of Theorem 1. Part of this hypothesis is that G_n be convex. Adding $b_n(y)$ to $h_n(y)$ has the effect of adding $b_n(y)$ to $G_n(y)$. This leads one to inquire as to whether $b_n(y)$ is convex. Fix D_n and notice, perhaps by drawing a picture, that $(D_n - y)^+$ is a convex function of y. Taking the expectation preserves convexity; hence, $b_n(y)$ is convex whenever $(R_n - \alpha R_{n+1})$ is nonnegative. In short, condition (a) of Theorem 1 holds whenever (b) to (h) hold, with

(h) $R_n \geq \alpha R_{n+1}$ for $n = 1, \ldots, N$

Condition (h) holds whenever the retail sales price has nonincreasing present value. It holds, for instance, when the interest rate is nonnegative and the retail sales price is constant. If (h) were violated, customers might balk rather than enter their orders into a backlog because they could face a price whose present value exceeds the price R_n they are currently willing to pay. Consequently, (h) seems to be a natural condition to include in the model.

The constant W_{N+1} has been selected in a way that causes $F(N + 1, \cdot)$ to be convex and nondecreasing. This suffices to establish hypothesis (d)' of Corollary 1; consequently, (h) suffices for the optimality of an (s, S)-policy.

Adding the convex term $b_n(\cdot)$ to $G_n(\cdot)$ preserves its convexity, but it need not preserve quasi-convexity. Hence, Theorem 1 adapts more readily than Theorem 2 to payment upon delivery because convexity is more robust than quasi-convexity.

If the following exercises, the hypothesis of Theorem 1 is seen to adapt to related variants of the model, but not to a third.

Exercise 7. In the case of payment of delivery, suppose that each customer whose order cannot be satisfied during the final period feels ill will equivalent to C dollars, where $C \geq e(1) - e(0)$. Take $W_{N+1} = C$, and adapt (7-15) and (7-6) to this situation. Show that Corollary 1 applies.

[**Hint:** First shows that $F(N + 1, \cdot)$ is convex and nondecreasing; then apply Lemma 5 of Supplement 2 to show that $EF(N, I_N(y, D_N))$ is convex.]

Exercise 8. In the case of payment upon delivery, suppose that the retailer is obligated to satisfy excess demand at the end of the planning horizon by paying setup cost K_{N+1}, purchasing at unit cost W_{N+1} and selling at unit price $R_{N+1} < W_{N+1}$. Adapt (7-15) and (7-16) to this situation. Show that Corollary 1 applies.

[**Hint:** Follow a pattern similar to that for Exercise 7, but aiming to use Lemma 3 of the final section of this chapter.]

Exercise 9. Suppose that customers pay when they receive their merchandise, but that they pay the price in effect when they ordered it. What goes wrong?

Backlogging with a Fixed Delivery Lag

Suppose that there is a fixed delay of λ periods between the placement of an order and its delivery. The order placed at period n has no effect on inventory until period $n + \lambda$, at which time it is received. Let us try for a dynamic programming formulation in which (n, x) is a state, this time with

$x =$ the quantity on hand *and* on order at the start of period n, just before placing that period's order

As before, $(y - x)$ is the order quantity; $K_n H(y - x) + (y - x)W_n$ is the ordering cost. Note, however, that the quantity on hand at period n is *not* included in the state; so the inventory carrying cost during period n cannot be deduced from (n, x). All units ordered prior to and at the start of period n will have been received by period $n + \lambda$ and, *in the case of backlogging*, the inventory at epoch $n + \lambda$ will be $y - \mathfrak{D}_n$, where

$$(7\text{-}17) \qquad \mathfrak{D}_n = D_n + D_{n+1} + \ldots + D_{n+\lambda-1}$$

So $h_{n+\lambda}(y - \mathfrak{D}_n)$ is the inventory carrying cost incurred during period $n + \lambda$. Theorems 1 and 2 remain valid when G_n is altered by replacing $h_n(y)$ by the discounted expect value of $h_{n+\lambda}(y - \mathfrak{D}_n)$, as follows.

Adaptation for delivery lag: Replace $h_n(y)$ by $\alpha^\lambda E h_{n+\lambda}(y - \mathfrak{D}_n)$.

Theorems 1 and 2 hypothesize, respectively, that h_n is convex and quasi-convex. Convexity of $h_n(y)$ assures convexity of $\mathbf{E}h_n(y - \mathfrak{D}_n)$, but quasi-convexity of $h_n(y)$ does not assure quasi-convexity of $\mathbf{E}h_n(y - \mathfrak{D}_n)$. Hence, the hypothesis of Theorem 1 is preserved when a fixed delay of λ periods is introduced, but the hypothesis of Theorem 2 need not be preserved.

The most interesting feature of the model having backlogging with a fixed lag in delivery is, perhaps, that the decision maker need only keep track of n and of x. Individual orders affect him, but not in ways subject to his control.

Exercise 10. Adapt Theorem 1 to the situation of backlogging with payment upon delivery and a fixed lag of λ periods between the placement of an order and the receipt of the merchandise.

Balking

Balking occurs when some customers accept backorders and others do not. We say a customer *balks* if she refuses to enter her order into a backlog or if she removes it from the backlog after having put it there. For the simplest model of balking, assume that 10% of the extant and the potential backlog balks in period n. The number of balkers is $0.1(D_n - y)^+$; hence, $I_n(y, D_n) = (y - D_n) + 0.1(D_n - y)^+$. Note, perhaps by drawing a picture, that, for each fixed value of D_n, the function $I_n(y, D_n)$ of y is continuous, convex, and non-decreasing. Hence, condition (c) of Theorems 1 and 2 holds for this model of balking. Divisibility is assumed here because $0.1(D_n - y)^+$ can be fractional-valued when y and D_n are integer-valued.

To model balking with indivisible merchandise, assume that each customer in the extant and the potential backlog acts independently and balks with probability p. To describe this situation, we introduce a set $\{Z_1, Z_2, \ldots\}$ of independent and identically distributed random variables, with $\Pr\{Z_i = 1\} = p$ and $\Pr\{Z_i = 0\} = 1 - p$ for $i = 1, 2, \ldots$. The number of balkers becomes $Z = Z_1 + \ldots + Z_T$ with $T = (D_n - y)^+$, and one gets

$$(7\text{-}18) \qquad I_n(y, D_n, Z) = (y - D_n) + \sum_{i=1}^{T} Z_i, \qquad \text{where } T = (D_n - y)^+$$

Exercise 11. Show that $I_n(y, D_n, Z)$ is, for each fixed D_n and Z, a convex nondecreasing function of y.

Yet another model of balking distinguishes customers who are asked to accept backlogged orders from those who have accepted backorders in earlier periods. Specifically, suppose that 80% of the customers who are asked to accept backorders do so, and suppose that no customers balk after accepting a backorder. The number of customers who balk equals $0.2(D_n - y^+)^+$; hence,

$$(7\text{-}19) \qquad\qquad I_n(y, D_n) = (y - D_n) + 0.2(D_n - y^+)^+$$

This expression for $I_n(y, D_n)$ is continuous and nondecreasing in y, but it is not convex. It satisfies the hypothesis of Theorem 2, but it violates the hypothesis of Theorem 1. Theorem 2 maintains this advantage when the customers who cannot be served immediately decide independently whether or not to balk.

Lost Sales and Deterioration of Stock

Deterioration of stock occurs when inventory is lost or stolen. Suppose that 5% of the inventory perishes (or disappears) at the end of the month, just after filling orders. In the lost-sales case, the inventory on hand at the start of the next month is given by $I_n(y, D_n) = 0.95(y - D_n)^+$. This function satisfies condition (c) of Theorems 1 and 2 [e.g., for each fixed value of D_n, $I_n(\cdot, D_n)$ is continuous, nondecreasing, and convex.]

However, (7-1) and (7-7) now account incorrectly for the sales receipts. The quantity $[y - I_n(y, D_n)]$ now equals the sum of the number of units sold at the retail sales price and the number of units that perish. The number of units that perish equals $0.05 I_n(y, D_n)$. To account correctly for the sales receipts, do as follows.

Adaptation for z% Deterioration of Stock: Add to $h_n(y)$ the term $0.01 z \alpha R_n E(y - D_n)^+$.

Note that this adaptation preserves the convexity of $h_n(y)$. Hence, the hypothesis of Theorem 1 remains valid when stock is allowed to deteriorate in this simple way. Divisibility is assumed because $0.05(y - D_n)^+$ can be fractional-valued when y and D_n are integer-valued. This model adapts to the case of indivisible merchandise in which each of the $(y - D_n)^+$ units spoils independently with probability p.

Exercise 12. Model the situation of payment upon delivery, 5% deterioration of stock, and 10% balking. Is the hypothesis of Theorem 1 preserved under this variant?

Backlogging and Deterioration of Stock

Suppose that 5% of the (positive) inventory perishes at the end of the month, just after filling orders. In the case of backlogging, $I_n(y, D_n) = (y - D_n) - 0.05(y - D_n)^+$. For each value of D_n, the function $I_n(\cdot, D_n)$ of y is now continuous, nondecreasing, and concave. In this case, $I_n(y, D_n)$ satisfies the hypothesis of Theorem 2, but it violates the hypothesis of Theorem 1. The prior adaptation still accounts correctly for the retail sales receipts. Theorem 2 applies if that adaptation yields a function $G_n(y)$ that is quasi-convex and whose minimizer a_n satisfies $a_1 \leq \ldots \leq a_N$.

SUMMARY

Inventory control models exist in many varieties. Stock may or may not deteriorate. Excess demand may be lost or backlogged. Backlogged customers may or may not balk. A customer may pay when he places an order or when he receives the merchandise. Customers whose orders are unfilled at the end of the planning horizon may benefit from a special order, or they may get their money back. A salvage market may exist for excess stock. The retail and wholesale prices may be fixed or may vary from period to period. A lag may occur between the time an order is placed with the wholesaler and the time the merchandise is delivered. Each of these features interacts with the others.

Nearly all these features give rise to inventory control models whose simplest and most representative forms satisfy the hypotheses of Theorems 1 and/or 2. The exception to this rule is a delivery lag. A fixed lag in delivery can be accommodated in conjunction with backlogging. Fixed and uncertain lags can be combined with other features *provided* that the manager is allowed to have at most one order outstanding at any time; see Problem 3.

The following, final section is devoted to the proof of Theorems 1 and 2.

PROOFS OF OPTIMALITY OF (s, S)-POLICIES*

The proof of Theorem 1 is accomplished by piecing together a quilt of four lemmas. The first of these lemmas is also used in the proof of Theorem 2.

One-Period Model

The line of analysis commences with the study of a one-period model. Throughout this discussion, K is a nonnegative constant, $H(\cdot)$ is the Heaviside function, and X is a subset of the real numbers. Interpret K and the setup cost and X as the set of allowable inventory levels. Let $L(\cdot)$ map X into the real numbers, and consider the solution $F(\cdot)$ to

$$(7\text{-}20) \qquad F(x) = \operatorname*{infimum}_{y \in X \mid y \geq x} \{KH(y - x) + L(y)\}, \qquad x \in X$$

The above is a single-period analogue of functional equation (7-8) and (7-9). We seek conditions under which an (s, S)-policy satisfies this equation.

With $K \geq 0$, the function L is called *K-quasi-convex* on X if there do not exist elements $a < b < c$ of X such that

$$(7\text{-}21) \qquad\qquad\qquad L(a) < L(b) > L(c) + K$$

Hence, an increase in a K-quasi-convex function cannot be followed by a decrease that exceeds K. Also, a 0-quasi-convex function is quasi-convex. Note that the function L_n in Figure 7-1 is K_n-quasi-convex.

LEMMA 1. Let X be a closed subset of the real numbers, and let the function L be lower semicontinuous and K-quasi-convex on X. Suppose, further, that X contains an element S such that

$$(7\text{-}22) \qquad\qquad\qquad L(S) = \inf \{L(x) \mid x \in X\}$$

---

* This section is advanced, and it uses Supplement 2.

Then (7-20) has the solution

$$(7\text{-}23) \qquad F(x) = \begin{cases} K + L(S), & x < s \\ L(x), & x \geq s \end{cases}$$

where s is given by

$$(7\text{-}24) \qquad s = \inf\{y \in X \,|\, L(y) \leq K + L(S)\}$$

Proof. Since $L(S) \leq K + L(S)$, (7-24) assures that $s \leq S$. Possibly, $s = -\infty$. Equation (7-23) will be verified in cases, depending on the position of x with respect to s and S.

Case 1: $x < s$. Equation (7-24) assures that $L(x) > K + L(S)$. Since $s \leq S$, the solution to (7-20) is $F(x) = K + L(S)$ in this case.

Case 2: $x = s > -\infty$. Since X is closed, $s \in X$. Hence, (7-24) and the lower semicontinuity of L assure that $L(x) \leq K + L(S)$, and the solution to (7-20) is $F(x) = L(x)$ in this case.

Case 3: $s < x \leq S$. Note from (7-24) that there exists $a \in X$ such that $s \leq a < x$ and $L(a) \leq K + L(S)$. Suppose that $L(x) > K + L(S)$. Then (7-21) is satisfied by a, $b = x$ and $c = S$. The K-quasi-convexity of L precludes this. Hence, $L(x) \leq K + L(S)$, and the solution to (7-20) is $F(x) = L(x)$ in this case.

Case 4: $x > S$. Suppose that $F(x) < L(x)$. Then (7-20) assures that there exists $c \in X$ that satisfies $c > x$ and $L(x) > K + L(c)$. Equation (7-22) assures that $L(S) \leq L(c)$. Hence, $L(S) < L(x)$. So (7-21) is satisfied by $a = S$, $b = x$, and c. This contradicts the K-quasi-convexity of L. Hence, $F(x) = L(x)$, which completes the proof. ∎

Multiperiod Model

To indicate why K-quasi-convexity does not suffice for the analysis of the two-period model ($N = 2$), we examine the simple (backlogging) case $I_1(y, D) = (y - D)$ for which (7-9) becomes

$$(7\text{-}25) \qquad L_1(y) = G_1(y) + \alpha E F(2, y - D_1)$$

Equation (7-25) indicates that the analysis of the multiperiod model is facilitated by conditions that are preserved under summation and under expectation. Quasi-convex and K-quasi-convex functions lack this property (e.g., the sum of two quasi-convex functions can have any number of local minima).

Sums of K-convex functions are, however, well behaved. Let L_1 and L_2 be, respectively, K_1 and K_2 convex on X. The following facts are immediate from the definition of K-convexity. The function $L_1 + L_2$ is $(K_1 + K_2)$-convex on X. If $\alpha > 0$, the function αL_1 is αK_1-convex on X. If $0 \leq \alpha \leq 1$, the function $\alpha L_1 + (1 - \alpha)L_2$ is $[\alpha K_1 + (1 - \alpha)K_2]$-convex on X. This extends to convex combinations of finitely many functions and, whenever the limits exist, to countable collections. Finally, if L is K-convex on the reals and if D is a real-valued random variable, $EL(x + D)$ is K-convex on the reals whenever the expectation exists. These properties distinguish K-convexity from K-quasi-convexity.

LEMMA 2.   A function L that is K-convex on X is K-quasi-convex on X.

Proof.   Suppose that a and b in X have $a < b$ and $L(a) < L(b)$. Since L is K-convex and $L(b) - L(a) > 0$, (7-14) assures that $L(c) + K > L(b)$ for all $c > b$. Hence, L is K-quasi-convex.   ■

On occasion, it proves convenient to manipulate the definition of K-convexity by writing $b = \alpha a + (1 - \alpha)c$, solving for α, and getting the following equivalent form of K-convexity; for $a < c$ and $0 < \alpha < 1$,

(7-26) $$L[\alpha a + (1 - \alpha)c] \leq \alpha L(a) + (1 - \alpha)[L(c) + K]$$

That 0-convexity is equivalent to convexity is immediate from the above.

Lemma 5 in Supplement 2 gives a condition under which a function of a function is convex. A condition under which a function of a function is K-convex is provided in

LEMMA 3.   Let h be convex and nondecreasing on X. Let g be K-convex on a set $Y \supseteq \{h(x) \mid x \in X\}$. If all elements $a < c$ of Y have

(7-27) $$g(a) \leq g(c) + K$$

then $g[h(x)]$ is K-convex on X.

Proof.   Consider elements $a < b < c$ of X. Take $\alpha = (c - b)/(c - a)$, so that $b = \alpha a + (1 - \alpha)c$. Note from (7-26) that K-convexity of gh is equivalent to nonnegativity of t, where

(7-28) $$t = \alpha gh(a) + (1 - \alpha)[gh(c) + K] - gh(b)$$

The hypothesis includes convexity of h; hence,

(7-29) $$h(b) \leq \alpha h(a) + (1 - \alpha)h(c)$$

The hypothesis that h is nondecreasing assures that $h(a) \leq h(b) \leq h(c)$, hence the existence of a number β between 0 and 1 such that

(7-30) $$h(b) = \beta h(a) + (1 - \beta)h(c)$$

Since g is K-convex on Y, (7-30) and (7-26) yield

(7-31) $$gh(b) \leq \beta gh(a) + (1 - \beta)[gh(c) + K]$$

Subtract (7-31) from (7-28) and collect terms to get

(7-32) $$t \geq (\beta - \alpha)[gh(c) + K - gh(a)]$$

We wish to show that $t \geq 0$. Since $h(a) \leq h(c)$, (7-27) assures that $[gh(c) + K - gh(a)] \geq 0$. Subtract (7-29) from (7-30) to get $0 \leq (\beta - \alpha)[h(c) - h(a)]$, hence $(\beta - \alpha) \geq 0$. So (7-32) assures that $t \geq 0$, hence gh is K-convex on X.   ■

The next lemma provides conditions under which F is a K-convex function that satisfies (7-27).

LEMMA 4.   Let X be a closed subset of the real numbers, and let L be a lower semicontinuous K-convex function on X. Then F is K-convex on X, and all elements $b < c$ of X have

(7-33) $F(b) \leq K + F(c)$

Proof.   Apply (7-20) with $x = b$ and $y = c$ to get $F(b) \leq K + F(c)$. So (7-33) holds. Lemma 2 shows that L is K-quasi-convex. Hence, Lemma 1 applies, and F is the solution to (7-23), with s given by (7-24). Consider elements $a < b < c$ of X, and note from (7-14) that K-convexity of F is equivalent to

(7-34) $F(b) \leq K + F(c) + \left(\dfrac{c-b}{b-a}\right)[F(a) - F(b)]$

If $F(a) \geq F(b)$, then (7-34) is immediate from (7-33). If $a \geq s$, Lemma 1 assures that $F(x) = L(x)$ for $x = a$, $x = b$, and $x = c$, and (7-34) is immediate from the K-convexity of L. In the remaining case, $a < s$ and $F(a) < F(b)$. Lemma 1 assures that $F(a) = F(s)$ and $s < b$. Hence, $b - a > b - s$, and

(7-35) $\left(\dfrac{c-b}{b-s}\right)[F(s) - F(b)] < \left(\dfrac{c-b}{b-a}\right)[F(a) - F(b)]$

Since (7-34) holds for $a = s$, (7-35) shows that (7-34) holds when $a < s$.   ■

Convexity and (s, S)-Policies

Convexity plays a central role in the hypothesis of Theorem 1, and the preceding lemmas are now molded into a proof of this theorem.

Proof of Theorem 1.   Condition (d) of the hypothesis shows that $F(N +1, \cdot)$ is a K_{N+1}-convex function that satisfies (7-33) with $K = K_{N+1}$ and that is lower semicontinuous.

This initializes the following inductive hypothesis: $F(n + 1, \cdot)$ is a K_{n+1}-convex function that is lower semicontinuous and that satisfies (7-33) with $K = K_{n+1}$. This and (c) let us apply Lemma 3 with $g(\cdot) = F(n + 1, \cdot)$ and $h(\cdot) = I_n(\cdot, D_n)$. Lemma 3 shows that $F[n + 1, I_n(\cdot, D_n)]$ is K_{n+1}-convex. Since (b) gives $K_n \geq \alpha K_{n+1}$, this shows that $\alpha F[n + 1, I_n(\cdot, D_n)]$ is K_n-convex. Since K-convexity is preserved under convex combinations, (e) suffices for the K_n-convexity of $\alpha E F[n + 1, I_n(y, D_n)]$. So (a) shows that $L_n(\cdot)$ is K_n-convex. That L_n is lower semicontinuous follows from lower semicontinuity of $F(n + 1, \cdot)$ and the continuity of $I_n(\cdot, D_n)$. Lemma 2 shows that $L_n(\cdot)$ is K_n-quasi-convex, and (7-22) follows from (e). Hence, Lemma 1 shows that $F(n, \cdot)$ satisfies (7-13). Lower semicontinuity of $F(n, \cdot)$ follows from (7-13) and from the lower semicontinuity of $L_n(\cdot)$. Also, Lemma 4 shows that $F(n, \cdot)$ is a K_n-convex function that satisfies (7-33) with $K = K_n$. This completes an inductive proof.   ■

Proof of Theorem 2.   Demonstration that L_n is K_n-quasi-convex is segmented into two claims.

Claim 1: Any elements b and c of X having $a_n < b < c$ also have $L_n(b) \leq K_n + L_n(c)$.

Proof of claim 1. Consider $x < z$. We first show that $F[n + 1, x] \leq K_{n+1} + F[n + 1, z]$. When $n = N$, this inequality is immediate from (d) and (7-6). When $n < N$, it follows directly from (7-5). Now consider $b < c$. Condition (c) assures that $I_n(b, D_n) \leq I_n(c, D_n)$; hence,

(7-36) $\mathbf{E}F[n + 1, I_n(b, D_n)] \leq K_{n+1} + \mathbf{E}F[n + 1, I_n(c, D_n)]$

for $n = 1, \ldots, N$. Substitute $y = b$ and $y = c$ in (7-9), subtract, and use (7-36) to get

$$L_n(b) - L_n(c) \leq \alpha K_{n+1} + G_n(b) - G_n(c)$$

Hence, (a) and (b) assure that $L_n(b) \leq K_n + L_n(c)$ when $a_n \leq b < c$. ∎

Claim 2: Any elements a and b of X having $a < b \leq a_n$ also have $L_n(a) \geq L_n(b)$.

Proof of claim 2. That $F[N+1, \cdot]$ is nonincreasing is immediate from (d) and (7-6). Consider $a < b$. That $F[N + 1, I_N(a, D_N)] \geq F[N + 1, I_N(b, D_N)]$ is then immediate from (c). So $\mathbf{E}F[N + 1, I_N(\cdot, D_N)]$ is nondecreasing. Hence (a) assures that $L_N(x)$ is nonincreasing for $x \leq a_N$.

Adopt the inductive hypothesis (just verified for the case $n = N - 1$) that $a < b \leq a_{n+1}$ assures that $L_{n+1}(a) \geq L_{n+1}(b)$. Consequently,

$$F(n + 1, a) = \min \{L_{n+1}(a), K_{n+1} + \inf_{y>a} \{L_{n+1}(y)\}$$

$$\geq \min \{L_{n+1}(b), K_{n+1} + \inf_{y>b} \{L_{n+1}(y)\} = F(n + 1, b)$$

So $F(n + 1, y)$ is nonincreasing for $y \leq a_{n+1}$. Hence, (c) assures that $\mathbf{E}F[n + 1, I_n(y, D_n)]$ is nonincreasing for $y \leq a_{n+1}$. Hence, (a) and (7-9) assure that $L_n(y)$ is nonincreasing for $y \leq a_n$. This completes an inductive proof of Claim 2.

To demonstrate that L_n is K_n-quasi-convex on X, we need to show that there exist no elements $a < b < c$ of X such that $L_n(a) < L_n(b) > K_n + L_n(c)$. Claims 1 and 2 show, respectively, that no such elements exist with $b \leq a_n$ and with $b \geq a_n$. Hence L_n is K_n-quasi-convex. Lower semicontinuity of L_n is easily verified. Hence, the existence of S_n satisfying (7-16) follows directly from (e). Consequently, Lemma 1 demonstrates the optimality of an (s, S)-policy. ∎

Exercise 13. Prove Corollary 1.
[Hint: Find a way to employ the proofs of Theorems 1 and 2.]

BIBLIOGRAPHIC NOTES

Scarf(1960) analyzed an inventory control model with set-up costs, backlogging, and convex operating costs. He introduced K-convexity and used it to show that an (s, S)-policy is optimal. Veinott (1966b) analyzed this model with quasi-convex operating costs; he obtained Theorem 2, though not quite in its current generality. Porteus (1971) introduced K-quasi-convexity. Tijms (1972) made further contributions. Schäl (1976) provided a line of development that unifies Theorems 1 and 2, but is rather complex. Lemma 3 is new, as are some of the adaptations in Section 4.

The early literature on inventory theory is insightfully surveyed by Scarf (1963) and Veinott (1966a).

PROBLEMS

1. (*Unsociable behavior*) Suppose, in (7-5) to (7-7), that $R_n = R$, $W_n = W$, $\alpha < 1$, $I_n(y, D_n) = (y - D_n)$, and $h_n(y) = (y)^+\alpha r W$. (The last of these conditions limits the holding cost to the investment in inventory.) Suppose that $2W > R$. What sort of policy is optimal for very large N?

2. (*Disposal of stock*) Suppose that the retailer can sell stock back to the wholesaler at the start of each period. In particular, suppose that W_n is the unit price for purchases from the wholesaler as well as for sell-backs to the wholesaler at the start of period n. Assume that the setup cost K_n is incurred on purchases, but not on shipbacks. For specificity, suppose that merchandise is indivisible and that excess demand is lost [i.e., $I_n(y, D_n) = (y - D_n)^+$] ~~lost-sales case~~
   a. Adapt functional equation (7-2) to this situation.
   b. Write a streamlined functional equation that has instantaneous transitions, one extra state, and an *average* of three decisions per state.
   c. Argue that an optimal policy for this model has one order-up-to quantity per period.
   [Hint: This is immediate from part b.]

   > *Remark:* If one could argue that it is not optimal to ship back merchandise once it is ordered, the optimality of an (s, S)-policy would follow directly from the streamlined formulation in part c of the problem.

3. (*Lag in delivery*) Let Λ denote the number of periods between the placement of an order with the supplier and the receipt of the merchandise. Adapt functional equation (7-2) to the following situations.
   a. $\Lambda = 1$.
   [Hint: Let x include the merchandise, if any, just received.]
   b. Λ is a random variable whose values are restricted to the set $\{0, 1\}$.
   c. Λ is a random variable whose values are restricted to the set $\{0, 1, 2\}$, and the inventory control manager is not allowed to place any order until all prior orders have been delivered.

4. (*Supplier out of stock*) At each period, the supplier is out of stock with probability $p > 0$. The supplier's resupply system is fast enough that the

probability he is out of stock in a given period is independent of the probability that he was out of stock in the preceding period. Also, your orders are so small a fraction of his business that their size does not affect the probability they are filled.

a. Alter (7-2) to reflect this.

b. Adapt the analysis to show that an (s, S)-policy is optimal.

5. Let $\mathbb{R}$ denote the set of real numbers, let $S \subseteq \mathbb{R}$ be an open interval, and let $F: S \longrightarrow \mathbb{R}$ be differentiable on S, with $f(x) = F'(x)$. Show that F is K-convex on S if and only if all elements $b < c$ of S have

$$F(c) + K \geq F(b) + (c - b)f(b)$$

6. List the uses made of $h_n(y)$ for the case $y \leq 0$.

7. Let X be a closed bounded subset of the real numbers, and let g be lower semicontinuous on X. Show that there exists $x \in X$ such that $g(x) \leq g(y)$ for all $y \in X$.

8. Prove or disprove: (1) the sum of two lower semicontinuous functions is lower semicontinuous; (2) the difference of two lower semicontinuous functions is lower semicontinuous.

INTRODUCTION

The preceding chapters provide repeated exposure to several of the themes that pervade dynamic programming. These themes include the roles played by states, functional equations, and the principle of optimality in the analysis of sequential decision processes. Several important themes have, however, not yet been introduced. For instance, no mention has yet been made of the roles played in dynamic programming by policy iteration, linear programming, or successive approximation. Nor has there been any mention of the so-called turnpike theorems that describe the limiting behavior of the finite-horizon model as the planning horizon becomes long.

These new themes are illustrated in this final chapter, which concerns a model of decision making in the face of an infinite planning horizon. The planning horizon is infinite, but a positive interest rate bounds the present value of the stream of income. This model is itself interesting, and it previews our companion volume, as it is the simplest of the models treated there.

THE MODEL

In our description of this model, the terms "period" and "epoch" preserve their special meanings. A *period* is an interval of time, and an *epoch* is the beginning of a period. A decision maker observes a system that evolves from state to state at equally spaced epochs that are numbered 0, 1, 2, . . . , without limit. At each epoch, the system can occupy one of finitely many *states* that are numbered

8

A DISCOUNTED

MARKOV

DECISION MODEL

Titles above in bold face type concern
advanced or specialized subjects. These
sections can be omitted without loss of
continuity.

1 through N, inclusive. At each epoch, the decision maker observes the current state of the system and selects a decision appropriate for it. Associated with each state i is a nonempty finite set $D(i)$ whose elements are called *decisions*. Whenever the decision maker observes state i, she must select some decision k from $D(i)$. If she observes state i and picks decision k, she immediately earns *reward* R_i^k, which may be negative. Transition from state to state is governed by chance; P_{ij}^k is the probability that the decision maker observes state j at the next epoch given that she now observes state i and selects decision k. Transitions are presumed to occur with probability of 1, which means that

$$(8\text{-}1) \qquad \sum_{j=1}^{N} P_{ij}^k = 1, \qquad \text{all } i, k$$

Should state j be observed at the next epoch, the decision maker must pick some decision k' from $D(j)$. This renews a cycle of events, which is repeated without limit. At each epoch, the decision maker faces a trade-off between maximizing her immediate reward (i.e., maximizing R_i^k over k) and her future rewards (i.e., getting transitions to states with large rewards).

This model is called a *discrete-time* model because decisions are made at evenly spaced epochs. It is called a *finite* model because it includes finitely many states and decisions. It is said to have an *infinite planning horizon* because decision making continues over an infinite period of time. It is called a *Markov* model because transition probabilities govern its evolution from state to state, as in a Markov chain. This model is called *stationary* because its data (the R_i^k's and the P_{ij}^k's) are independent of the number n of periods that have elapsed since decision making began. These data are presumed to be known to the decision maker; if the transition probabilities were not known precisely, the model would be called *adaptive* or *Bayesian*.

Every period in this model has the same length, and the model includes a per period *interest rate* r. Consequently, the decision maker is presumed to be indifferent between receiving X units of income at epoch n and $\alpha^n X$ units of income at epoch 0, where $\alpha = 1/(1 + r)$. The interest rate r is presumed to be positive, hence the *discount factor* α satisfies

$$(8\text{-}2) \qquad 0 < \alpha < 1$$

This model is called *discounted* because α satisfies (8-2).

A *policy* δ is taken, as usual, as a decision procedure that specifies for each state i a decision $\delta(i)$ in $D(i)$. To *use* policy δ is to select decision $\delta(i)$ at each epoch at which state i is observed. The *policy space* Δ is the set of all alternative policies. Every combination of decisions is allowed; that is, δ is in Δ if and only if $\delta(i)$ is in $D(i)$ for $i = 1, 2, \ldots, N$. The cognoscenti will recognize that attention is limited here to *stationary* policies; for instance, the decision $\delta(i)$ for state i can depend on i, but not on the number n of periods that have elapsed since decision making began.

As the model is finite, it has smallest reward m and largest reward M, where[1]

(8-3) $$m = \min_{i,k} \{R_i^k\}, \qquad M = \max_{i,k} \{R_i^k\}$$

The income (reward) that the decision maker receives at any given epoch lies between m and M, inclusive. Hence, the present value at epoch 0 of the income she receives at epoch n lies between $m\alpha^n$ and $M\alpha^n$. Similarly, the present value at epoch 0 of the entire stream of income lies between $m/(1 - \alpha)$ and $M/(1 - \alpha)$; for instance, the present value is no more than the sum, $M/(1 - \alpha)$, of the geometric series $M + M\alpha + M\alpha^2 + \ldots$.

Decision making commences at epoch 0 with the observation of some state i and the selection of some decision k for it. This determines the reward R_i^k earned at epoch 0. The state observed at epoch 1 is influenced by chance; hence the reward that will be earned at epoch 1 is a random variable, one whose value lies between m and M. Similarly, the present value at epoch 0 of the entire stream of income is a random variable whose value lies between $m/(1 - \alpha)$ and $M/(1 - \alpha)$. The decision maker is presumed to act so as to maximize the expectation of the present value at epoch 0 of the stream of income (rewards) she receives. We recall from Chapter 6 that an individual who maximizes the expectation of his or her income is said to have a *linear utility function*.

The model is, at least, completely specified. It is a stationary finite discounted discrete-time Markov decision model with an infinite planning horizon and a linear utility function.

This is the simplest of the models that are treated in our companion volume, whose content can now be previewed. Some of its models allow the time between decision-making epochs to be variable or uncertain. Some of its models allow that number of states and/or decisions to be infinite. Some of its models include the possibility of stopping as well as of continuing to make decisions. Some of its models have interest rate of zero; (i.e., $\alpha = 1$). Some of its models have growth rates; effectively, $\alpha > 1$. Some of its models have nonlinear utility functions. In some of its models, the decision maker does not know the transition probabilities, but learns about them as decision making proceeds. In some of its models, evolution is governed by transition rates rather than by probabilities. Some of its models include two or more decision makers, either in concert or in opposition. All of its models are analyzed by generalizations of the techniques introduced here.

MAIN RESULTS

The main results of this chapter are stated here and proved subsequently. We recall that to use policy δ is to choose decision $\delta(i)$ at each epoch at which state i is observed. Vector notation is introduced now because it will prove convenient.

---

[1] When the range of a variable is omitted, the entire range is indicated (i.e., m is the smallest R_i^k over all i and k).

Associated with each policy δ is an $N \times 1$ vector v^δ whose ith component v_i^δ is the expectation of the present value at epoch 0 of the income stream that is received if state i is observed at epoch 0 and if policy δ is used. The quantity v_i^δ is abbreviated to the *expected discounted income* earned by starting at state i and using policy δ.

Similarly, the *optimal return vector* f is the $N \times 1$ vector whose ith element f_i is defined by

$$(8\text{-}4) \qquad f_i = \max_{\delta \in \Delta} \{v_i^\delta\}, \qquad i = 1, \ldots, N$$

The model has finitely many states and decisions. So it has finitely many policies, and the maximum on the right-hand side of (8-4) is attained.

Exercise 1. Recall why f_i is bounded, with $m/(1 - \alpha) \leq f_i \leq M/(1 - \alpha)$.

If the decision maker observes state i at epoch 0, she wishes to choose a policy π such that $v_i^\pi = f_i$. The first main result of this chapter is that there exists a policy π such that

$$(8\text{-}5) \qquad f_i = v_i^\pi, \qquad i = 1, \ldots, N$$

A policy π that satisfies (8-5) is called an *optimal* policy. An optimal policy must attain N maxima simultaneously, so the existence of an optimal policy is not trivial. Its existence cannot be demonstrated by induction "backward" from the end of the planning horizon to the beginning, as this planning horizon lacks an end.

The second main result of this chapter is that the optimal return vector f is the unique solution to the system of equations

$$(8\text{-}6) \qquad f_i = \max_{k \in D(i)} \{R_i^k + \sum_{j=1}^{N} \alpha P_{ij}^k f_j\}, \qquad i = 1, \ldots, N$$

That is, the vector f defined by (8-5) solves (8-6), and f is the only solution to (8-6). Recognize (8-6) as a *functional equation* of dynamic programming.

We shall also see how to find an optimal policy from the solution to functional equation (8-6). For each state i and decision k, define s_i^k by

$$(8\text{-}7) \qquad s_i^k = f_i - R_i^k - \sum_{j=1}^{N} \alpha P_{ij}^k f_j$$

Note from (8-6) that each s_i^k is nonnegative and that $0 = \min_k \{s_i^k\}$. Set

$$(8\text{-}8) \qquad D^*(i) = \{k \in D(i): s_i^k = 0\}, \qquad i = 1, \ldots, N$$

Hence, $D^*(i)$ contains k if and only if k maximizes the right-hand side of (8-6). Set

$$(8\text{-}9) \qquad \Delta^* = \{\delta \in \Delta: \delta(i) \in D^*(i) \text{ for } i = 1, \ldots, N\}$$

Hence, Δ^* contains the policies whose decisions maximize the right-hand sides of (8-6). We shall see that Δ^* is the set of optimal policies; that is, $v^\delta = f$ if and only if $\delta \in \Delta^*$. Ties in (8-6) seem unlikely; one would normally expect Δ^* to consist of exactly one policy π, which would then be the lone optimal policy.

To find f, one could enumerate all the policies in Δ. Since the number of elements of Δ is the product, over i, of the number of elements of $D(i)$, the set Δ can contain astronomically many elements, even when the numbers of states and decisions per state are modest. The third main result of this chapter is that f can be found from the solution to the linear program:

Program 1. Minimize $\left\{\sum_{i=1}^{N} w_i\right\}$, subject to the constraints

(8-10)
$$w_i - \sum_{j=1}^{N} \alpha P_{ij}^k w_j \geq R_i^k, \qquad \text{all } i, k$$

$$w_i \text{ unrestricted}, \qquad\qquad \text{all } i$$

We shall see that Program 1 is feasible and bounded and that its objective is minimized, uniquely, by $w = f$.

Exercise 2. Show how to find Δ^* from the solution to Program 1.

The fourth main result of this chapter is that f can be computed by the method known as policy iteration. This method consists of alternating "policy evaluation" and "policy improvement" steps whose exact specifications are postponed to a later section. We also see that this particular policy iteration scheme is equivalent, pivot for pivot, to the application of the dual simplex method to Program 1. Since linear programming codes are widely available, it would be pointless to code this particular policy iteration scheme. Nonetheless, policy iteration is a fundamental idea in dynamic programming. We shall, for instance, use policy iteration to demonstrate that a solution exists to functional equation (8-6). In addition, certain other sequential decision processes have solutions that can be found by policy iteration but not (at least not yet) by linear programming or its extensions.

The chapter's fifth main result concerns a method for approximating the optimal return vector, f. Starting with any $N \times 1$ vector v^0, one computes $N \times 1$ vectors v^1, v^2, and so on, iteratively from the equation

(8-11)
$$v_i^n = \max_{k \in D(i)} \{R_i^k + \sum_{j=1}^{N} \alpha P_{ij}^k v_j^{n-1}\}, \qquad i = 1, \ldots, N$$

With v^0 on the right-hand side of (8-11) one computes v_i^1 for each i. Then, with v^1 on the right-hand side of (8-11), one computes v_i^2 for each i, and so forth. Iterative computations of this sort embody the method of *successive approximation*, which is sometimes called *value iteration*. We shall see that v^n converges to f; that is,

(8-12)
$$f_i = \lim_{n \to \infty} \{v_i^n\}, \qquad i = 1, \ldots, N$$

This chapter also contains two variants of the successive approximation scheme given as (8-11), a trick for accelerating (8-11), and a heuristic for combining (8-11) with its variants. By contrast with linear programming, successive approximation avoids the need to solve systems of equations. But successive approximation does not compute f exactly.

The sixth "main result" is actually a family of results that relate long, finite planning horizons to infinite planning horizons. Members of this family are called *turnpike theorems*. One member is already in evidence. Recognize (8-11) as a functional equation that is consistent with the following interpretation: v_i^n is the largest attainable expected discounted income when one starts at state i and faces an n-epoch planning horizon with terminal reward vector v^0. With this interpretation, (8-12) shows that, as n becomes large, the optimal return vector v^n for the n-epoch planning horizon approaches the optimal return vector f for the infinite-horizon model. So (8-12) is a simple turnpike theorem.

A deeper and more important turnpike theorem relates the optimal policies for the n-epoch and infinite-period models. We shall see that there exists a threshold n_0 such that one can use policies in Δ^* whenever the number n of epochs from now to the end of the planning horizon exceeds n_0. Hence, if Δ^* consists of the single policy π, one gets

$$v_i^n = R_i^{\pi(i)} + \sum_{j=1}^{N} \alpha P_{ij}^{\pi(i)} v_j^{n-1}, \qquad i = 1, \ldots, N$$

for all $n \geq n_0$. In this case, one can use (the optimal) policy π until the final n_0 epochs of the planning horizon.

The seventh main result is that stationary policies suffice in the following sense. One cannot increase f by introducing policies whose decisions are allowed to depend on the prior history of the process. This result may seem trite, but there *do* exist sequential decision processes with stationary data but for which stationary policies are not sufficient. The proof that stationary policies suffice for this particular model is deferred to the companion volume, where the requisite tools will be at hand.

POLICY EVALUATION

The first six main results are proved in logical order, not quite in numerical order. We start by suppressing the optimization problem and studying the properties of a particular policy. The superscript k on R_i^k and P_{ij}^k is suppressed, temporarily, and the optimization problem simplifies to the following Markov process with rewards.

A system is observed at epochs $0, 1, \ldots$, with limit. At each epoch, the system occupies one of N states that are numbered 1 through N. If state i is observed at some epoch, reward R_i is earned then, and state j will be observed at the next epoch with probability P_{ij}. As in (8-1), the probabilities sum to 1; that is, $\sum_{j=1}^{N} P_{ij} = 1$ for $i = 1, \ldots, N$. The interest rate r remains positive, and the discount factor α still satisfies $0 < \alpha < 1$.

Let R be the $N \times 1$ vector whose ith element is R_i, let P be the $N \times N$ matrix whose ijth element is P_{ij}, and let Q be the $N \times N$ matrix whose ijth element equals αP_{ij}; that is,

$$Q = \alpha P$$

Suppose for the moment that state i is observed at epoch 0. The probability that state j will be observed at epoch 1 is P_{ij}. The conditional probability law will soon be used to justify the equation

(8-13) $$\Pr \{\text{state } j \text{ is observed at epoch 2}\} = \sum_{k=1}^{N} P_{ik} P_{kj}$$

That is, the probability that state j will be observed at epoch 2 is the sum over k of the probability P_{ik} that state k will be observed at epoch 1 times the conditional probability P_{kj} that state j will be observed at epoch 2 if state k is observed at epoch 1. It is clear from (8-13) that the probability of observing state j at epoch 2 is the ijth element of the matrix $P^2 = P \cdot P$. Similarly, the probability that state j will be observed at epoch 3 is the ijth element of the matrix P^3.

Still on the assumption that state i is observed at epoch 0, we list in Table 8-1 the expectations of the rewards received at various epochs. Needless

TABLE 8-1. Expected reward earned at epoch m if state i is observed initially

Epoch, m	Expectation of the Reward Earned at Epoch m	Its Present Value at Epoch 0
0	R_i	R_i
1	$\sum_{j=1}^{N} P_{ij} R_j$	$\alpha \sum_{j=1}^{N} P_{ij} R_j$
2	$\sum_{j=1}^{N} (\sum_{k=1}^{N} P_{ik} P_{kj}) R_j$	$\alpha^2 \sum_{j=1}^{N} \sum_{k=1}^{N} P_{ij} P_{kj} R_j$
m	ith element of $P^m R$	ith element of $Q^m R$

to say, R_i is earned immediately. The expectation of the reward earned at epoch 1 is the sum over j of the probability P_{ij} of a transition to state j times the reward R_j earned if that transition occurs. The present value at epoch 0 of the reward earned at epoch 1 is obtained by multiplying this reward by α. Similarly, the expectation of the reward earned at epoch 2 is the sum over j of the probability $(P^2)_{ij}$ of a transition to state j times the reward R_j earned if state j is observed. Its present value at epoch 0 is obtained by multiplying it by α^2. The general situation should now be evident; the present value at epoch 0 of the expectation of the reward earned at epoch m is the ith element of the vector $Q^m R = (\alpha P)^m R$. The preceding expression makes clear the value of matrix notation. For $n = 0, 1, \ldots$, let $V(n)$ be the $N \times 1$ vector whose ith element has the following interpretation.

$$V(n)_i = \text{the expectation of the present value at epoch 0 of}$$
the income earned during epochs 0 through $n - 1$
if state i is observed at epoch 0

The expectation of the sum of random variables is the sum of their expectations, whether or not they are independent. Hence, it follows from Table 8-1 that

(8-14) $$V(n) = R + QR + \ldots + Q^{n-1}R$$

Our main concern in the infinite-horizon model lies in the limiting behavior of $V(n)$ as n approaches infinity.

Exercise 3. Justify (8-15) below.

(8-15) $$V(n) = R + QV(n - 1)$$

Equations (8-14) and (8-15) give further testimony to the usefulness of matrix notation. Let I denote the $N \times N$ identity matrix, and factor R in (8-14) to get

(8-16) $$V(n) = (I + Q + \ldots + Q^{n-1})R$$

Equation (8-16) makes it clear that the behavior of $V(n)$ as n approaches infinity is governed by the limiting behavior of the sum of the first n powers of the matrix $Q = \alpha P$. A square matrix, like P, whose entries are nonnegative and whose rows sum to 1 is called a *stochastic* matrix.

THEOREM 1. Let P be a stochastic matrix and let $Q = \alpha P$, where $0 < \alpha < 1$. Then $(I - Q)$ is invertible, and

(8-17) $$(I - Q)^{-1} = I + Q + Q^2 + \ldots$$

Proof. From $Q = \alpha P$, we know that the entries of Q are nonnegative, and so are the entries of Q^n. Since the rows of P sum to 1, the rows of $Q^n = \alpha^n P^n$ sum to α^n. Consequently, the entries of Q^n approach 0 as $n \to \infty$.

Cancel intermediary terms to verify that

(8-18) $$(I - Q)(I + Q + \ldots + Q^{n-1}) = I - Q^n$$

Consider a row vector x for which $x(I - Q) = 0$. Premultiply (8-18) by x to see that $0 = x - xQ^n$, hence $x = xQ^n$. Let $n \to \infty$, recall that $Q^n \to 0$, and conclude that $x = 0$. Hence, $(I - Q)$ is invertible. Premultiply (8-18) by $(I - Q)^{-1}$ to get $I + \ldots + Q^{n-1} = (I - Q)^{-1} - (I - Q)^{-1}Q^n$. Let $n \to \infty$; since $Q^n \to 0$, this verifies (8-17), completing the proof. ∎

Theorem 1 applies to the sum $(I + Q + \ldots + Q^{n-1})$ in (8-16). So $V(n)$ converges as n approaches infinity, and this entitles us to define the $N \times 1$ vector v by

(8-19) $$v = \lim_{n \to \infty} \{V(n)\} = (I - Q)^{-1}R$$

Interpret v_i as the total expected discounted income for starting at state i; that is, v_i is the expectation of the present value at epoch 0 of the stream of income earned over the infinite planning horizon, provided that state i is observed at epoch 0.

When comparing two $N \times 1$ vectors w and x, we write $w \geq x$ if $w_i \geq x_i$ for $i = 1, \ldots, N$. Similarly, we write $w \leq x$ if $w_i \leq x_i$ for $i = 1, \ldots, N$. Theorem 2 provides conditions under which $v \geq w$ [where v is defined by (8-19)] and under which $v \leq w$.

THEOREM 2.   Let P be a stochastic matrix; let $Q = \alpha P$, where $0 < \alpha < 1$; and let $v = (I - Q)^{-1}R$. Consider any $N \times 1$ vector w.

(a) If $R + Qw \geq w$, then $v \geq R + Qw$.

(b) If $R + Qw \leq w$, then $v \leq R + Qw$.

Remark: Interpret R as a reward vector, $Q = \alpha P$ as a transition matrix, and $R + Qw$ as the reward vector for the truncated discounted one-transition process in which one receives terminal reward w_j if transition occurs to state j. Part (a) asserts that if this truncated process improves on w (i.e., if $R + Qw \geq w$), then a further improvement is obtained by using reward vector R and transition matrix Q indefinitely (i.e., $v \geq R + Qw$).

Proof.   Define the $N \times 1$ vector t by

(8-20) $$t = R + Qw - w = R - (I - Q)w$$

The hypothesis of part (a) is that $t \geq 0$. Premultiply (8-20) by $(I - Q)^{-1}$ and then use (8-19) to get

(8-21) $$(I - Q)^{-1}t = v - w$$

Since $t \geq 0$ and $Q \geq 0$, (8-17) assures that

(8-22) $$(I - Q)^{-1}t = \sum_{i=0}^{\infty} Q^i t \geq Q^0 t = t$$

Combine (8-21) and (8-22) into $v \geq w + t$, then use (8-20) to get $v \geq R + Qw$. This proves part (a). For part (b), reverse all inequalities.   ■

OPTIMIZATION

The decision problem reemerges now. That is, k reappears as a superscript on R_i^k, P_{ij}^k, and $Q_{ij}^k = \alpha P_{ij}^k$. Also, δ appears on the $N \times 1$ vector R^δ and on the $N \times N$ matrix P^δ, where the ith element of R^δ equals $R_i^{\delta(i)}$ and where the ijth element of P^δ equals $P_{ij}^{\delta(i)}$. Similarly, δ appears as a superscript on the vectors $V^\delta(n)$ and v^δ and on the $N \times N$ matrix Q^δ, where

(8-23) $$Q^\delta = \alpha P^\delta$$

Since P^δ is stochastic and since $0 < \alpha < 1$, the matrix $(I - Q^\delta)$ is invertible, and (8-19) gives

(8-24) $$v^\delta = (I - Q^\delta)^{-1}R^\delta$$

Interpret v_i^δ as the expected discounted income for starting in state i and using policy δ. Now that v_i^δ is precisely specified by (8-24), the optimal return vector f is precisely specified by (8-4). Premultiply (8-24) by $(I - Q^\delta)$ to see that v^δ is the unique solution to the equation

$$(8\text{-}25) \qquad\qquad v^\delta = R^\delta + Q^\delta v^\delta$$

Equation (8-25) is known as the *policy evaluation* equation.

POLICY ITERATION

Although policy iteration appeared fourth on the list of "main results," it is described before the others because we shall use it to demonstrate that an optimal policy exists. For our Markov decision model, the policy iteration algorithm is as follows.

1. (*Initialization*) Pick any policy δ in Δ.
2. (*Policy evaluation*) Set $t_i = 0$ for $i = 1, \ldots, N$, and compute $v^\delta = (I - Q^\delta)^{-1} R^\delta$.
3. (*Policy improvement*) DO for $i = 1, \ldots, N$.
    4. DO for each k in $D(i)$.
        5. Set $t_i^k = R_i^k + \sum_{j=1}^{N} Q_{ij}^k v_j^\delta - v_i^\delta$.
        6. If $t_i^k > t_i$, set $\delta(i) \leftarrow k$ and $t_i \leftarrow t_i^k$.
7. Stop if $t_i = 0$ for $i = 1, \ldots, N$. Otherwise, go to step 2.

Remark: We recall that DO loops are nested by indentation [e.g., step 4 executes step 5 and then 6 for each k in $D(i)$]. Also, the symbol $\leftarrow$ means replacement [e.g., $\delta(i) \leftarrow k$ replaces the value of $\delta(i)$ by the value k].

Computational remark: At least one decision changes at each nonterminal iteration of policy iteration. If only a few decisions change at a particular iteration, the matrix requiring inversion differs by only a few rows from the one whose inverse had been calculated. In that case, it is more efficient to update the old inverse than to reinvert from scratch, just as is the case in linear programming. Indeed, the inverse can even be stored in product form.

Recall from (8-5) that an optimal policy π must attain N maxima simultaneously. The next theorem uses policy iteration to construct such a policy.

THEOREM 3. The policy iteration routine terminates finitely, and the last policy π that is evaluated in step 2 has $v^\pi = f$.

Proof. In comparing vectors w and x, we write $w > x$ if $w \geq x$, with $w_i > x_i$ for at least one i. Similarly, $w < x$ if $w \leq x$, with $w_i < x_i$ for at least one i.

Consider any nonterminal iteration of step 7. Let π be the policy evaluated in the preceding execution of step 2, and let δ be the policy to be evaluated in the next execution of step 2. Array the current values of t_1 through t_N into the

$N \times 1$ vector t, and note that

$$(8\text{-}26) \qquad\qquad 0 < t = R^\delta + Q^\delta v^\pi - v^\pi$$

Apply part (a) of Theorem 2 with $w = v^\pi$, $R = R^\delta$, and $Q = Q^\delta$ to get $v^\delta \geq R^\delta + Q^\delta v^\pi > v^\pi$. Hence, $v^\delta > v^\pi$, and each nonterminal iteration of the policy evaluation step improves on all preceding steps. There are finitely many policies, hence the policy iteration routine must terminate in finitely many steps. Let π be the last policy evaluated; since policy iteration terminates with each t_i^k nonpositive, the inequality

$$(8\text{-}27) \qquad\qquad 0 \geq R^\delta + Q^\delta v^\pi - v^\pi$$

holds for every policy δ. Apply part (b) of Theorem 2 with $w = v^\pi$, $R = R^\delta$, and $Q = Q^\delta$ to get $v^\delta \leq v^\pi$. Since this holds for every policy δ, one gets $v^\pi = f$; hence, policy π is optimal. ∎

A FUNCTIONAL EQUATION

The first main result has just been verified. The second main result is that the optimal return vector f is the unique solution to functional equation (8-6), which is repeated below.

$$(8\text{-}6) \quad f_i = \max_k \left\{ R_i^k + \sum_{j=1}^N \alpha P_{ij}^k f_j \right\}, \qquad i = 1, \ldots, N$$

THEOREM 4. The optimal return vector f satisfies (8-6), and no other vector satisfies (8-6).

   Proof. To show that f satisfies (8-6), we examine the conditions that cause policy iteration to terminate. Policy iteration terminates when it finds a policy π having, for each i, $0 = t_i^{\pi(i)} = \max_k \{t_i^k\}$. Theorem 3 shows that $v^\pi = f$; hence, the current values of the t_i^k's give

$$(8\text{-}28) \qquad 0 = \max_k \left\{ R_i^k + \sum_{j=1}^N Q_{ij}^k f_j - f_i \right\}, \qquad i = 1, \ldots, N$$

Add f_i to both sides of (8-28) to transform it to (8-6). Hence, f satisfies (8-6).

   Now consider any vector g that satisfies (8-6). For each state i, let $\pi(i)$ be a decision that maximizes the right-hand side of (8-6), so that

$$(8\text{-}29) \qquad\qquad g = R^\pi + Q^\pi g$$

Also, for every policy δ,

$$(8\text{-}30) \qquad\qquad g \geq R^\delta + Q^\delta g$$

Theorem 2 shows that $g = v^\pi$ and that $g \geq v^\delta$ for every δ. Hence, $g = \max_\delta \{v^\delta\} = f$, which completes a proof of Theorem 4. ∎

   The set Δ^* is defined by (8-7) to (8-9). Verification that Δ^* is the set of optimal policies is left as an exercise.

Exercise 4. Show that the set Δ^* is nonempty and that π is in Δ^* if and only if $v^\pi = f$.

LINEAR PROGRAMMING

The third "main result" is that f can be found by solving a linear program, namely Program 1.

THEOREM 5. f is feasible for Program 1, and the objective of Program 1 is minimized, uniquely, by f.

Proof. It is immediate from Theorem 4 that f is feasible for Program 1. Theorem 3 shows that there exists an optimal policy π. Consider any feasible solution w to Program 1. The constraints in Program 1 that correspond to policy π form the vector inequality $w - Q^\pi w \geq R^\pi$. Part (b) of Theorem 2 shows that $w \geq v^\pi = f$. Hence, the objective, $\sum_{i=1}^N w_i$, is minimized by f and only by f. ■

Exercise 5. Show how to find an optimal policy from a solution to Program 1.

LINEAR PROGRAMMING IS POLICY ITERATION

The point of this section is that linear programming and policy iteration are essentially the same [i.e., policy iteration amounts to multiple substitution (block pivoting) in the simplex method]. Skillfully programmed simplex codes are widely available; there is no point in coding the policy iteration routine previously discussed. Here and hereafter, $\mathbf{1}$ is the $N \times 1$ vector of 1's, and $\mathbf{1}^t$ is the $1 \times N$ vector of 1's.

Almost all of this section presumes familiarity with linear programming, and you may skip to its final remark without loss of continuity. We shall investigate the dual of Program 1, which is written below.

Program 2.[2]   Maximize $\{\sum_i \sum_k x_i^k R_i^k\}$, subject to the contsraints

(8-31)
$$\sum_k x_i^k - \sum_j \sum_k x_j^k Q_{ji}^k = 1, \qquad \text{all } i$$

$$x_i^k \geq 0, \qquad \text{all } i, k$$

The following theorem shows that the feasible bases for Program 2 are in one-to-one correspondence with the policies.

THEOREM 6. A matrix B is a feasible basis for Program 2 if and only if $B = I - Q^\delta$ for some policy δ.

---

[2] When a range of summation is omitted, the entire range of the variable is indicated; e.g., the objective in Program 2 is the sum of $x_i^k R_i^k$ over all pairs (i, k) with $i = 1, \ldots, N$ and k in $D(i)$.

Proof. First let δ be any policy, and define the $1 \times N$ vector x^δ by

(8-32) $x^\delta = 1'(I - Q^\delta)^{-1}$

Thus, as $(I - Q)^{-1} \geq 0$ [see (8-17)], one has

(8-33) $x^\delta(I - Q^\delta) = 1', \qquad x^\delta \geq 0$

Compare the above with the constraints for Program 2; with the other variables set equal to zero, x^δ is feasible. Since $(I - Q^\delta)$ is invertible, $B = (I - Q^\delta)$ has full rank. Hence, B is a feasible basis.

Now consider any feasible basis B for Program 2, and let the row vector x be its basic feasible solution (i.e., $xB = 1'$, and $x \geq 0$). Since there are N constraints in (8-31), there are at most N positive elements in the row vector x. But (8-31) also assures that

$$\sum_k x_i^k \geq 1, \qquad i = 1, \ldots, N$$

Hence, there are exactly N positive elements in x. Moreover, for every i, there exists exactly one k for which x_i^k is positive. These positive variables identify a policy δ, with $B = (I - Q^\delta)$. ∎

In order to examine a pivot step on linear programming, we first define the $1 \times N$ vector A_i^k.

(8-34) $A_i^k = (-Q_{i1}^k, \ldots, 1 - Q_{ii}^k, \ldots, -Q_{iN}^k)$

Constraint (8-31) of Program 2 can now be written in row-vector notation, as

$$\sum_i \sum_k x_i^k A_i^k = 1'$$

where $1'$ is (still) the $1 \times N$ vector of 1's.

THEOREM 7. Consider the application of the simplex routine to Program 2. It makes the same sequence of pivots as does policy iteration, when the latter changes at each policy improvement step only the one decision for which t_i^k is most positive.

Proof. Consider a pivot step. Theorem 6 indicates that the current basis corresponds to some policy, say δ. The current basis is $(I - Q^\delta)$. In a pivot, the row that enters the basis is the one for which

$$R_i^k - A_i^k(I - Q^\delta)^{-1}R^\delta$$

is most positive. (This is the row-notation analogue of $c_j - c_B B^{-1}A_j$.) Substitute (8-24) and then (8-34) into the above, getting

(8-35) $R_i^k - A_i^k(I - Q^\delta)^{-1}R^\delta = R_i^k - A_i^k v^\delta = R_i^k + \sum_j Q_{ij}^\delta v_j^\delta - v_i^\delta$

The extreme right-hand side of (8-35) is identical to t_i^k, as defined in policy iteration! So the row chosen to enter the basis is the one for which t_i^k is most positive. Should row A_i^k enter, row $A_i^{\delta(i)}$ must leave, as a feasible solution must have, for each i, exactly one x_i^k positive. This completes the proof of Theorem 7. ■

One might hope that computation is speeded up by the sort of multiple substitution that occurs in policy iteration. However, most large-scale simplex codes garner this advantage automatically. These codes have a "pricing pass" in which all columns whose prices are favorable (e.g., t_i^k positive) are ranked and the best dozen (say) are noted. The pricing pass alternates with a "pivot pass" in which the noted columns are brought into the basis one by one as long as their prices remain favorable. Except for the recheck of prices, this is close to what policy iteration does.

It is now well known that the following two routines make precisely the same sequence of pivots: the simplex routine applied to any linear program and the dual simplex routine applied to its dual. This fact and Theorem 7 assure us that policy iteration is the same as block pivoting in the dual simplex method, as applied to Program 1.

> *Remark:* Theorem 5 and its proof remain valid if the objective of Program 1 is changed from $\mathbf{1}'w$ to cw, where c is any $1 \times N$ vector having $c_i > 0$ for each i. (Check this.) Also, each basis $(I - Q^\delta)$ for Program 2 has a nonnegative inverse, which means that Program 2 is a *Leontief substitution system*. In fact, we just observed a key property of Leontief substitution systems, which is that a fixed basis remains optimal for all positive right-hand sides.

SUCCESSIVE APPROXIMATION

Successive approximation consists of starting with an arbitrary $N \times 1$ vector v^0 and computing $N \times 1$ vectors v^1, v^2, and so on, iteratively from (8-11), which is reproduced below.

$$\textbf{(8-11)} \quad v_i^n = \max_k \left\{ R_i^k + \sum_{j=1}^{N} \alpha P_{ij}^k v_j^{n-1} \right\}, \qquad i = 1, \dots, N$$

The fifth main result is that v^n approaches f as $n \to \infty$. To prove this, we take as the norm of an $N \times 1$ vector u the largest of the absolute values of its entries:

$$\textbf{(8-36)} \qquad \qquad \|u\| = \max_i \{|u_i|\}$$

The following theorem shows that v^n is closer to f (in this norm) than is v^{n-1}.

THEOREM 8. For $n = 1, 2, \dots,$

$$\textbf{(8-37)} \qquad \qquad \|v^n - f\| \le \alpha \|v^{n-1} - f\|$$

Proof. Let k maximize the right-hand side of (8-11):

$$\textbf{(8-38)} \qquad \qquad v_i^n = R_i^k + \sum_{j=1}^{N} \alpha P_{ij}^k v_j^{n-1}$$

The functional equation (8-6) gives

(8-39)
$$f_i \geq R_i^k + \sum_{j=1}^{N} \alpha P_{ij}^k f_j$$

Subtract (8-39) from (8-38), and collect terms to get

$$v_i^n - f_i \leq \sum_{j=1}^{N} \alpha P_{ij}^k (v_j^{n-1} - f_j)$$

Note from (8-36) that

$$(v_j^{n-1} - f_j) \leq |v_j^{n-1} - f_j| \leq ||v^{n-1} - f||$$

Hence,

(8-40)
$$v_i^n - f_i \leq \sum_{j=1}^{N} \alpha P_{ij}^k ||v^{n-1} - f|| = \alpha ||v^{n-1} - f||$$

Now let l maximize the right-hand side of (8-6):

(8-41)
$$f_i = R_i^l + \sum_{j=1}^{N} \alpha P_{ij}^l f_j$$

Note from (8-11) that

(8-42)
$$v_i^n \geq R_i^l + \sum_{j=1}^{N} \alpha P_{ij}^l v_j^{n-1}$$

Subtract (8-42) from (8-41) and repeat the preceding analysis to get

(8-43)
$$f_i - v_i^n \leq \alpha ||f - v^{n-1}||$$

Combine (8-40) and (8-43) to get

$$|v_i^n - f_i| \leq \alpha ||v^{n-1} - f||$$

Then maximize over i to verify (8-37), which completes the proof of Theorem 8. ∎

Repeated application of Theorem 8 gives

(8-44)
$$||v^n - f|| \leq \alpha^n ||v^0 - f||$$

In other words,

(8-45)
$$|v_i^n - f_i| \leq \alpha^n ||v^0 - f||, \qquad i = 1, \ldots, N$$

Since $\alpha^n \rightarrow 0$ as $n \rightarrow \infty$, inequality (8-45) shows that $v_i^n \rightarrow f_i$ as $n \rightarrow \infty$. This verifies (8-12).

Exercise 6. Suppose that $v^0 = 0$. Recall that $m = \min_{i,k} \{R_i^k\}$ and that $M = \max_{i,k} \{R_i^k\}$. Show that $||v^n - f|| \leq \max \{|m|, |M|\} \alpha^n / (1 - \alpha)$.

SPEED OF SUCCESSIVE APPROXIMATION
AND OF LINEAR PROGRAMMING

There are two key measures of the speed of a computer program. One is the number of computer operations (memory accesses, arithmetic operations, etc.) needed to execute it. The other is the number of cells of fast-access memory needed to execute it without substantial compromise in speed. Both measures are important because they limit the size of the problem that can be solved efficiently on a given computer.

This section compares the speed of computation of f by the method of successive approximation and by linear programming. One parameter in this comparison is the average number K of decisions per state. We shall count multiplications rather than computer operations. This is justified because both methods entail numbers of memory accesses, additions, and comparisons that are comparable to their respective numbers of multiplications. Readers who are willing to accept Table 8-2 on faith should skip the next three technical paragraphs.

TABLE 8-2. Measures of speed of computation of f by successive approximation and by linear programming

	Successive Approximation	Linear Programming
Number of fast-access cells for data	N	N^2
Number of multiplications per iteration	KN^2	$(K + 1)N^2$
Number of iterations	Infinite	$3KN$, in practice

One column of Table 8-2 relates to successive approximation. The other relates to linear programming, specifically to the application of the dual simplex method to Program 1. The top row counts the number of fast-access memory cells needed for data storage; one can make do with fewer cells than this, but at the expense of substantial time spent moving data from slow-access memory (e.g., tape) to fast-access memory (e.g., chips) and back. The middle row counts multiplications per iteration. The bottom row counts iterations.

Successive approximation consists of repeated application of (8-11). The N elements of v^{n-1} are kept in fast-access memory, as these data are used in each execution of (8-11). The rewards and transition probabilities for particular state-decision pairs need not be kept in fast-access memory; they can read sequentially, if necessary. The number of multiplications needed to compute v_i^n from (8-11) is, on average, KN, and N such computations must be executed per iteration. The number of iterations is infinite, as this is an approximate method.

The dual simplex method, as applied to Program 1, is virtually identical to policy iteration, but with a flexible rule about which of the positive t_i^k's to pivot in. The number of elements of fast-access storage is shown as N^2, the number of elements in the inverse $(I - Q^\delta)^{-1}$ of the current basis. When, however, the inverse is kept in product form, it is *not* necessary to have all factors in the product in core at once [e.g., the vector $(I - Q^\delta)^{-1}R^\delta$ can be formed by applying the products to R^δ sequentially]. Even so, we accept N^2 as representative of the number of fast-access elements needed to store the data. The rewards and transition probabilities for the nonbasic columns need not be kept in fast-access memory; they can be read sequentially, if necessary. The number of multiplications needed to update the basis is N^2, as is the number of multiplications needed to compute $v^\delta = (I - Q^\delta)^{-1}R^\delta$. The number of multiplications needed to execute a complete pricing pass is $(K - 1)N^2$, as this amounts to computing t_i^k for each of the $(K - 1)N$ nonbasic state-decision pairs. This accounts for the three addends in the total $(K + 1)N^2$. The number of pivots is shown as $3KN$ because, in practice, the simplex method takes roughly three pivots per variable.

As Table 8-2 indicates, the primary advantage of successive approximation over linear programming is its reduced requirement for fast-access storage. When the number N of states is in the tens of thousands, this advantage is telling.

To do sensitivity analysis, one needs optimal solutions to the primal and dual problems. We have seen how to approximate an optimal solution to Program 1 by successive approximation. To approximate a solution to Program 2, take as δ a policy whose decisions maximize the right-hand side of (8-11) at the final iteration. With $\mathbf{1}^t$ as the $1 \times N$ vector of 1's, one wishes a solution x^δ to $x^\delta(I - Q^\delta) = \mathbf{1}^t$ or, equivalently, to $x^\delta = \mathbf{1}^t + x^\delta Q^\delta$. To approximate x^δ, iterate

$$x^n = \mathbf{1}^t + x^{n-1}Q^\delta$$

That x^n converges to x^δ as $n \rightarrow \infty$ is a routine consequence of the fact that $(Q^\delta)^n$ converges to 0 as $n \rightarrow \infty$.

NUMERICAL DIFFICULTIES

When the epochs are closely spaced, the per epoch interest rate r is very small, and the discount factor α is very close to 1. In policy iteration and in linear programming, one needs to solve for v^δ the policy evaluation equation,

$$(8\text{-}46) \qquad\qquad (I - \alpha P^\delta)v^\delta = R^\delta$$

With $\mathbf{1}$ as the $N \times 1$ vector of 1's, we note that

$$(8\text{-}47) \qquad\qquad (I - \alpha P^\delta)\mathbf{1} = (1 - \alpha)\mathbf{1}$$

Consequently, $(1 - \alpha)$ is an eigenvalue of $(I - \alpha P^\delta)$. This means that the determinant of the matrix $(I - \alpha P^\delta)$ approaches 0 as α approaches 1. For α near 1, the matrix $(I - \alpha P^\delta)$ is very nearly singular, and numerical difficulties arise when one attempts to solve (8-46). In a linear programming code, round-off error will accumulate rapidly as the inverse is updated, making frequent reinversions necessary.

When α is close to 1, numerical difficulties also afflict successive approximation. The quantity α^n approaches 0 very slowly. Hence, convergence is slow, and numerical errors are very slowly attenuated.

The undiscounted model studied in the companion to this volume provides a useful surrogate for the discounted model when the discount factor is close to 1, and the undiscounted model is not afflicted by these numerical difficulties.

TURNPIKE THEOREMS

An infinite planning horizon is often, if not always, a surrogate for a finite planning horizon that is long and, perhaps, of somewhat uncertain length. To relate the infinite-horizon model to a finite-horizon model, interpret (8-11) as a functional equation. In particular, with

$$(8\text{-}48) \qquad\qquad v^0 = 0$$

interpret v^n as follows.

$v_i^n =$ expected discounted income for starting in state i and operating the system optimally over an n-epoch planning horizon

Since $v^0 = 0$, inequality (8-45) simplifies to

$$(8\text{-}49) \qquad\qquad |v_i^n - f_i| \leq \alpha^n \| f \|$$

Consequently, as n becomes large, the optimal return v_i^n for starting in state i in the n-epoch model approaches the optimal return f_i for starting in state i in the model with an infinite planning horizon. This is a simple turnpike theorem.

It is not normally the case that v^n is attained by a (stationary) policy whose decisions ignore n; for instance, in the final period, the decision maker should maximize his immediate reward without regard to the terminal state of the system, as $v^0 = 0$.

Define the sequence $(\pi^1, \pi^2, \ldots)$ of policies so that, for each n, policy π^n maximizes the right-hand side of (8-11). Consequently,

$$(8\text{-}50) \qquad v^n = R^{\pi^n} + \alpha P^{\pi^n} v^{n-1}, \qquad n = 1, 2, \ldots$$

Note that it is optimal to use policy π^n when n is the number of epochs from now to the *end* of the planning horizon.

We recall that the Δ^*, as defined by (8-7) to (8-9), is the set of optimal policies for the infinite-horizon model. We shall soon see that π^n is in Δ^* for all n that exceed some threshold n_0.

Recall from (8-6) and (8-7) that each s_i^k is nonnegative, and take c as the smallest positive s_i^k. That is

$$(8\text{-}51) \qquad\qquad c = \min \{s_i^k : s_i^k > 0\}$$

The quantity c appears in the following bound on n_0.

THEOREM 9. Assume that $v^0 = 0$. Then policy π^n is in Δ^* for all $n > n_0$, where

$$(8\text{-}52) \qquad\qquad n_0 = \frac{\log c - \log \|2f\|}{\log \alpha}$$

Remark: It is tacitly assumed that at least one s_i^k is positive. If each s_i^k were zero, one would have $\Delta^* = \Delta$, and Theorem 9 would be true (but dull) with $n_0 = 0$.

Proof. Equation (8-52) is independent of the base chosen for the logarithm. Choose any base greater than 1, so that $\log \alpha < 0$. Consider any $n > n_0$, so (8-52) gives

$$(8\text{-}53) \qquad n \log \alpha < \log c - \log \|2f\| = \log\left[\frac{c}{\|2f\|}\right]$$

Exponentiation preserves strict inequalities:

$$(8\text{-}54) \qquad\qquad \alpha^n < \frac{c}{\|2f\|}$$

Consider any $n > n_0$ and any i. Set $k = \pi^n(i)$. To prove the theorem, we must show that $s_i^k = 0$. Add (8-11) to (8-7) to get

$$(8\text{-}55) \qquad s_i^k = (f_i - v_i^n) + \sum_{j=1}^{N} \alpha P_{ij}^k (v_j^{n-1} - f_j)$$

Inequality (8-49) gives $|f_i - v_i^n| \le \alpha^n \|f\|$ and (8-49) also gives $|v_j^{n-1} - f_j| \le \alpha^{n-1} \|f\|$. Substitute these into (8-55), getting

$$(8\text{-}56) \qquad s_i^k \le \alpha^n \|f\| + \alpha^{n-1} \|f\| \sum_{j=1}^{N} \alpha P_{ij}^k = 2\alpha^n \|f\|$$

the last since $\sum_j P_{ij}^k = 1$.

Finally, rearrange (8-54) as $2\alpha^n \|f\| < c$, and substitute into (8-56), getting $s_i^k < c$. Since each s_i^k is nonnegative, (8-51) assures that $s_i^k = 0$. This completes the proof. ■

When, as seems normal, $\Delta^* = \{\pi\}$, Theorem 9 states that it is optimal to use policy π for all but the final n_0 epochs of a long, finite planning horizon. That is the strongest sort of turnpike theorem one can hope for.

Theorem 9 is often helpful in the analysis of infinite-horizon models. One can often show, by induction on n, that a structured policy is optimal for the n-epoch model. Since π^n is in Δ^* for all sufficiently large n, an optimal policy for the infinite-horizon model must inherit this structure. For instance, if an (s_n, S_n)-policy is optimal for the finite-horizon model, Theorem 9 shows that an (s, S)-policy is optimal for the infinite-horizon model. We must note that inductive arguments of this sort usually hinge on adroit selection of v^0 (usually with $v^0 \ne 0$). Hence, (8-48) needs to be relaxed, the result of which is to increase the bound n_0 in Theorem 9.

The bound on n_0 in Theorem 9 is based on worst-case analysis. This bound has $\log \alpha$ in the denominator, and $\log \alpha$ is unpleasantly close to 0 when α is close to 1. The bound on n_0 can, however, be tightened slightly by a more delicate analysis, the result of which is to replace $\| 2f \|$ by $(\max_i f_i - \min_j f_j)$.

ACCELERATING SUCCESSIVE APPROXIMATION*

The successive approximation method presented earlier is but one of three well-known iterative methods for approximating the optimal return function f. Two other methods are presented here, together with a trick for speeding up the original method.

In preparation, we describe successive approximation more abstractly. Let $\mathbb{R}^N$ denote the set of all $N \times 1$ vectors having real entries. A function E whose domain is $\mathbb{R}^N$ and whose range is contained in $\mathbb{R}^N$ is called an *operator* on $\mathbb{R}^N$. An element x of $\mathbb{R}^N$ is called a *fixed point* of E if $x = E(x)$. Some operators on $\mathbb{R}^N$ have no fixed points. Others have several.

One often wishes to find a fixed point of an operator E on $\mathbb{R}^N$. One method that sometimes works is to apply E repeatedly. This method is known as *successive approximation*; it consists of picking a vector x^0 in $\mathbb{R}^N$ and generating vectors x^1, x^2, and so on, by

$$(8\text{-}57) \qquad x^n = E(x^{n-1}), \qquad n = 1, 2, \ldots$$

Even if an operator has exactly one fixed point, x^n need not approach it as $n \to \infty$. With this preamble, we limit our attention to operators E that satisfy Condition (*), below.

CONDITION (*). There exists a number β, with $0 < \beta < 1$, such that

$$(8\text{-}58) \qquad E(u) \leq E(v) + \beta c\mathbf{1}$$

for all u and v in $\mathbb{R}^N$ and all numbers $c \geq 0$ such that $u \leq v + c\mathbf{1}$. (We recall that $\mathbf{1}$ is the $N \times 1$ vector of 1's.)

When $c = 0$, Condition (*) is such that $u \leq v$ assures that $E(u) \leq E(v)$. Condition (*) may seem contrived, but it holds for the successive approximation schemes of interest, and it yields the following theorem.

THEOREM 10. Suppose that the operator E on $\mathbb{R}^N$ satisfies Condition (*). Then

  a. There exists exactly one $x = E(x)$.
  b. For any x^0 in $\mathbb{R}^N$, the sequence $\{x^n : n = 0, 1, \ldots\}$ defined by (8-57) satisfies

$$(8\text{-}59) \qquad x = \lim_{n \to \infty} \{x^n\}$$

---

* This advanced section employs more complicated mathematics. It may be omitted without loss of continuity.

c. With $\|\cdot\|$ defined on $\mathbb{R}^N$ by (8-36), one has

(8-60) $\|x - E(y)\| \leq \|E(y) - y\| \dfrac{\beta}{1 - \beta}$,   all $y \in \mathbb{R}^N$

d. If $E(y) \geq y$, then $x \geq E(y)$. If $E(y) \leq y$, then $x \leq E(y)$.

Remark: Part (a) is that x is the unique fixed point of E. Part (b) is that x^n converges to x as $n \longrightarrow \infty$. Part (c) uses the information, y and $E(y)$, obtained from a single application of E to bound the distance $\|x - E(y)\|$ between $E(y)$ and the fixed point, x. Part (d) resembles Theorem 2; it gives conditions under which $x \geq y$ and under which $x \leq y$.

Note: The following proof of Theorem 10 uses contraction mappings and should be skipped by readers who are unfamiliar with them.

Proof.   For any x in $\mathbb{R}^N$, let $(x)^+ = \max_i \{\max\{0, x_i\}\}$. If x has any positive elements, $(x)^+$ is the largest positive element; otherwise, $(x)^+ = 0$. Fix u and v in $\mathbb{R}^N$. Note that

(8-61) $u \leq v + (u - v)^+ \mathbf{1}$, $v \leq u + (v - u)^+ \mathbf{1}$

So Condition (*) gives

(8-62) $E(u) \leq E(v) + \beta(u - v)^+ \mathbf{1}$, $E(v) \leq E(u) + \beta(v - u)^+ \mathbf{1}$

Since $\|u - v\| = \max\{(u - v)^+, (v - u)^+\}$, inequality (8-62) assures that

(8-63) $\|E(u) - E(v)\| \leq \beta \|u - v\|$

Inequality (8-63) shows that E is a contraction mapping on the complete metric space $(\mathbb{R}^N, \rho)$, where $\rho(u, v) = \|u - v\|$. Parts (a) and (b) are the fixed-point theorem for contraction mappings; see Kolmogorov and Fomin (1957, p. 43). Part (c) follows from parts (a) and (b); we omit its proof. For the first half of part (d), suppose that $y \leq E(y)$. Use (8-57) with $x^0 = y$. We argue by induction that $x^0 \leq \ldots \leq x^n$, which is true by hypothesis when $n = 1$. Suppose that it is true for n. Then $x^{n-1} \leq x^n = E(x^{n-1})$, and Condition (*) is satisfied with $u = x^{n-1}$, $v = x^n$, and $c = 0$. Hence, $E(x^{n-1}) \leq E(x^n)$ (i.e., $x^n \leq x^{n+1}$). Since $x^n \longrightarrow x$, this assures that $x^n \leq x$ and, in particular, that $x^1 = E(y) \leq x$. The other half of part (d) is proved similarly.   ∎

We shall see that Condition (*) is satisfied by the following three methods of successive approximation.

Method 1

For each policy δ, define the operator H_δ on $\mathbb{R}^N$ by

(8-64) $[H_\delta(u)]_i = R_i^{\delta(i)} + \alpha \sum\limits_{j=1}^{N} P_{ij}^{\delta(i)} u_j$, $i = 1, \ldots, N$

Also, define the operator A on $\mathbb{R}^N$ by

(8-65) $[A(u)]_i = \max\limits_{\delta \in \Delta} \{[H_\delta(u)]_i\}$, $i = 1, \ldots, N$

Successive approximation method 1 consists of repeated application of A. That is, with u^0 as a fixed vector in $\mathbb{R}^N$, one computes u^1, u^2, and so on, from

(8-66) $$u^n = A(u^{n-1}), \qquad n = 1, 2, \dots$$

Compare (8-64) to (8-66) with (8-11) to see that method 1 is the one studied previously (i.e., if $u^0 = v^0$, then $u^n = v^n$ for each n). Even so, we state and prove the following lemma.

LEMMA 1. For each policy δ, the operator H_δ defined by (8-64) satisfies Condition (*), and so does A, each with $\beta = \alpha$.

Proof. Let $u, v \in \mathbb{R}^N$ and $c \geq 0$ satisfy $u \leq v + c\mathbf{1}$. Since $P_{ij}^k \geq 0$ and $\sum_j P_{ij}^k = 1$, we have, for each i,

$$[H_\delta(u)]_i = R_i^{\delta(i)} + \alpha \sum_{j=1}^N P_{ij}^{\delta(i)} u_j$$

$$\leq R_i^{\delta(i)} + \alpha \sum_{j=1}^N P_{ij}^{\delta(i)}(v_j + c)$$

$$= [H_\delta(v)]_i + \alpha c$$

The above holds for each i; hence,

(8-67) $$H_\delta(u) \leq H_\delta(v) + \alpha c\mathbf{1}$$

Since $\alpha < 1$, inequality (8-67) verifies Condition (*) with $\alpha = \beta$. Maximize the right-hand side of (8-67) over δ to get $H_\delta(u) \leq A(v) + \alpha c\mathbf{1}$; then maximize the left-hand side over δ to get $A(u) \leq A(v) + \alpha c\mathbf{1}$. Hence, A satisfies Condition (*), proving the lemma. ∎

Lemma 1 shows that Theorem 10 applies to H_δ and to A, with $\beta = \alpha$ in both cases. In particular, A has a unique fixed point, which is easily seen to be f. Theorem 10 also shows that u^n converges to f, that $\|u^n - f\| \leq \|u^n - u^{n-1}\| \cdot \beta/(1 - \beta)$, that $f \geq u^n$ if $u^n \geq u^{n-1}$, and that $f \leq u^n$ if $u^n \leq u^{n-1}$.

Exercise 7. Let δ satisfy $u^n = H_\delta(u^{n-1})$, where u^n is defined by (8-66). Use Theorem 10 twice to show that $\|v^\delta - f\| \leq 2\|u^n - u^{n-1}\| \alpha/(1 - \alpha)$. Can this bound be tightened when $u^n \geq u^{n-1}$?

Exercise 8. Use the operators H_δ and A to write policy iteration succinctly.

Method 2

In (8-11), every component of v^n is computed with v^{n-1} on the right-hand side. This fails to use the latest information; it is simpler and perhaps faster to replace v_i^{n-1} by v_i^n on the right-hand side of (8-11) as soon as the latter is computed, rather than at the end of the iteration. This motivates the following redefinition of the operators H_δ and A; for $i = 1, \dots, N$,

$$(8\text{-}68) \qquad [H_\delta(u)]_i = R_i^{\delta(i)} + \alpha \sum_{j=1}^{i-1} P_{ij}^{\delta(i)}[H_\delta(u)]_j + \alpha \sum_{j=i}^{N} P_{ij}^{\delta(i)}u_j$$

$$(8\text{-}69) \qquad [A(u)]_i = \max_k \{R_i^k + \alpha \sum_{j=1}^{i-1} P_{ij}^k [A(u)]_j + \alpha \sum_{j=i}^{N} P_{ij}^k u_j\}$$

The new operators H_δ and A differ from the old ones in that, for $j < i$, the quantities $[H_\delta(u)]_j$ and $[A(u)]_j$ replace u_j. Hence, the new values get used when (8-68) and (8-69) are executed for each i in numerical order.

The computations obtained with H_δ and A defined by (8-68) and (8-69) are called *method 2*. Since A changes, so does the sequence $u^n = A(u^{n-1})$. Equations (8-68) and (8-69) have transition probabilities that sum to 1. But, with foresight, we allow $\sum_j P_{ij}^k < 1$ in the following lemma.

LEMMA 2. Relax (8-1) to $\sum_j P_{ij}^k = S_i^k \leq 1$ for each i and k, and set $S = \max_{i,k} \{S_i^k\}$. Then, for every δ, the operator H_δ defined by (8-68) satisfies Condition (*), and so does the operator A defined by (8-69), each with $\beta = \alpha S$.

Proof. Let $u, v \in \mathbb{R}^N$ and $c \geq 0$ satisfy $u \leq v + c\mathbf{1}$. Direct substitution into (8-68) gives $[H_\delta(u)]_1 \leq [H_\delta(v)]_1 + \alpha S_1^{\delta(1)}c$. Adopt the inductive hypothesis, just verified for the case $i = 2$, that $[H_\delta(u)]_j \leq [H_\delta(v)]_j + \alpha Sc$ for all $j < i$. From $0 \leq S \leq 1$ and $0 \leq \alpha c$, one gets $\alpha Sc \leq \alpha c$; hence, $[H_\delta(u)]_j \leq [H_\delta(v)]_j + \alpha c$ for $j < i$. Substitute this into (8-68), together with $u_j \leq v_j + c$ for $j \geq i$, to get

$$[H_\delta(u)]_i \leq R_i^{\delta(i)} + \alpha \sum_{j=1}^{i-1} P_{ij}^{\delta(i)}\{[H_\delta(v)]_j + \alpha c\} + \alpha \sum_{j=i}^{N} P_{ij}^{\delta(i)}\{v_j + c\}$$

From $c > 0$ and $0 < \alpha < 1$, we get $\alpha^2 c < \alpha c$. This inequality, the one displayed above, and (8-68) give

$$[H_\delta(u)]_i \leq [H_\delta(v)]_i + \alpha c S_i^{\delta(i)}$$

Since $\alpha c S_i^{\delta(i)} \leq \alpha c S$, the inequality displayed above completes an inductive proof that H_δ satisfies Condition (*) with $\beta = \alpha S$. Proof that A satisfies Condition (*) with $\beta = \alpha S$ is omitted because it follows exactly the same pattern as for H_δ. ∎

Lemma 2 shows that Theorem 10 applies to the operators H_δ and A in method 2. In particular, A has a unique fixed point, which is easily seen to be f. Also, u^n converges to f as $n \to \infty$.

Exercise 9. Show that v^δ and f are the unique fixed points of the operators H_δ and A in method 2. Show that the results in Exercise 7 apply in this case.

Method 3

The idea of using the latest information is driven to the limit by the following method. When computing $[H_\delta(u)]$ in increasing i, use the new values $[H_\delta(u)]_j$ for $j < i$, the old values u_j for $j > i$, and solve for $[H_\delta(u)]_i$, getting

(8-70) $\quad [H_\delta(u)]_i = \{R_i^{\delta(i)} + \alpha \sum_{j=1}^{i-1} P_{ij}^{\delta(i)}[H_\delta(u)]_j + \alpha \sum_{j=i+1}^{N} P_{ij}^{\delta(i)} u_j\}/[1 - \alpha P_{ii}^{\delta(i)}]$

By the same token, the operator A becomes

(8-71) $\quad [A(u)]_i = \max_k \{R_i^k + \alpha \sum_{j=1}^{i-1} P_{ij}^k [A(u)]_j + \alpha \sum_{j=i+1}^{N} P_{ij}^k u_j\}/[1 - \alpha P_{ij}^k]\}$

The computations obtained with H_δ and A defined by (8-70) and (8-71) are called *method 3*.

LEMMA 3. For every policy δ, the operator H_δ defined by (8-70) satisfies Condition (*), and so does the operator A defined by (8-71), each with

(8-72) $$\beta = \alpha \max_{i,k} \left(\frac{1 - P_{ii}^k}{1 - \alpha P_{ii}^k} \right)$$

Remark: From $0 \leq P_{ii}^k \leq 1$ and $0 < \alpha < 1$, one gets $\beta \leq \alpha$.

Proof. Equations (8-70) and (8-71) suggest the following reinterpretation of the data. The reward for observing state i and selecting decision k becomes $R_i^k/(1 - \alpha P_{ii}^k)$, not R_i^k. For $i \neq j$, the transition probability becomes $P_{ij}^k/(1 - \alpha P_{ii}^k)$, not P_{ij}^k. For $i = j$, the transition probability becomes 0. The revised transition probabilities sum to $S_i^k = (1 - P_{ii}^k)/(1 - \alpha P_{ii}^k)$. Moreover, (8-70) and (8-71) with the original data are identical to (8-68) and (8-69) with the revised data. Hence, Lemma 2 proves Lemma 3. ∎

Lemma 3 shows that Theorem 10 applies to the operators H_δ and A in method 3. One can check that their fixed points are v^δ and f, respectively, hence that u^n converges to f.

Exercise 10. Show that v^δ and f are, respectively, the fixed points of H_δ and of A in method 3.

Gauss and Seidel are often credited with using the latest information in successive approximation, Jacobi with solving for x_i with the other variables fixed. One might then call method 2 a Gauss–Seidel method and method 3 a Gauss–Seidel–Jacobi method. Although these methods have advantages over method 1, they lack its ability to exploit jumps.

Method 1 with Jumps

We now return to method 1, in which H_δ is defined by (8-64). Note that for any scalar c,

(8-73) $$H_\delta(u + c\mathbf{1}) = H_\delta(u) + \alpha c\mathbf{1}$$

Equation (8-73) rests on (8-64); it is *not* satisfied by the definitions of H_δ in methods 2 and 3. Method 1 is now accelerated by allowing jumps.

1. Pick any u^0 in R^N. Set $n = 0$.
2. Calculate $A(u^n)$.

3. Set $z^n = 0$. If $A(u^n) \geq u^n$, set $z^n = \min_i \{A(u^n)_i - u^n_i\}$. If $A(u^n) \leq u^n$, set $z^n = \max_i \{A(u^n)_i - u^n_i\}$. Set

$$(8\text{-}74) \qquad\qquad u^{n+1} = A(u^n) + \frac{1z^n\alpha}{1-\alpha}$$

4. Let $n \leftarrow (n+1)$. Go to step 2.

At an iteration n having $z^n = 0$, one gets $u^{n+1} = A(u^n)$, as before. When $z^n \neq 0$, equation (8-74) gives a "jump" of $z^n\alpha/(1-\alpha)$ in each element of u^{n+1}. One gets $z^n > 0$ when $u^n_i < A(u^n)_i$ for each i, and one gets $z^n < 0$ when $u^n_i > A(u^n)_i$ for each i. These jumps are analyzed in Lemma 4.

LEMMA 4. Method 1 with jumps has this property.

    (a) Suppose that $u^n \leq A(u^n)$. Then, for every $r \geq n$, $u^r \leq A(u^r)$, $z^r \geq 0$, and $u^r \geq u^{r+1}$.

    (b) Suppose that $u^n \geq A(u^n)$. Then, for every $r \geq n$, $u^r \geq A(u^r)$, $z^r \leq 0$, and $u^r \geq u^{r+1}$.

Remark: Suppose that $z^n > 0$ [i.e., $A(u^n)_i > u^n_i$ for each i]. Part (a) asserts that $z^r \geq 0$ for all $r > n$ (i.e., an "up-jump" cannot be followed by "down-jumps"). Part (a) also asserts that $u^r \leq A(u^r)$, which combines with part (d) of Theorem 10 to show that $u^r \leq f$ for each $r \geq n$. Hence, these "up-jumps" cannot "overshoot" f. Finally, one can show that if $z^n > 0$, then method 1 with jumps has, for each $r > n$, u^r that lies strictly above (hence closer to f) the u^r obtained by method 1 without jumps. Similar remarks hold when $z^n < 0$.

Proof. We prove part (a), as part (b) follows the same pattern. By hypothesis, $u^n \leq A(u^n)$. Adopt the inductive hypothesis that $u^r \leq A(u^r)$. That $z^r \geq 0$ is immediate from its definition, as is the inequality

$$A(u^r) \geq u^r + 1z^r$$

Combine the above with the definition (8-74) of u^{r+1} to get

$$(8\text{-}75) \qquad\qquad u^{r+1} \geq u^r + 1z^r + \frac{1z^r\alpha}{1-\alpha} = u^r + \frac{1z^r}{1-\alpha}$$

Hence, $u^{r+1} \geq u^r$. Also, Condition (*) applies to (8-75) with $c = 0$, $u = u^r + 1z^r/(1-\alpha)$ and $v = u^{r+1}$; hence,

$$(8\text{-}76) \qquad\qquad A(u^{r+1}) \geq A\left(u^r + \frac{1z^r}{1-\alpha}\right)$$

Equation (8-73) holds for every δ; hence, $A(u + c1) = A(u) + \alpha c1$. Combine this with (8-76) to get

$$(8\text{-}77) \qquad\qquad A(u^{r+1}) \geq A(u^r) + \frac{1z^r\alpha}{1-\alpha} = u^{r+1}$$

the last from the definition (8-74) of u^{r+1}. Inequality (8-77) shows that $A(u^{r+1}) \geq u^{r+1}$. This completes the inductive step. Since we have also shown that $z^r \geq 0$ and $u^r \leq u^{r+1}$, it completes the proof of part (a). Proof of part (b) is omitted because it follows precisely the same pattern. ∎

An Edge for Methods 2 and 3

Method 1 enjoys the advantage just presented. However, methods 2 and 3 enjoy an advantage when $u^n \leq A(u^n)$, that is detailed in the following exercise.

Exercise 11. For this exercise, switch notation as follows: while (8-64) remains the same, H_δ is replaced by J_δ in (8-68) and by K_δ in (8-70).

a. Suppose that $H_\delta(u) \geq u$. Show that $J_\delta(u) \geq u$.

b. Suppose that $J_\delta(u) \geq u$. Show that $K_\delta(u) \geq u$.

c. Make and prove similar statements for the three versions of the one-stage optimization operator A.

d. Why might part (c) fail to hold with its inequalities reversed?

A Heuristic

Exercise 11 suggests that (when maximizing) one prefers $A(u^n) \geq u^n$ to $A(u^n) \leq u^n$. This suggests the following heuristic. Start with u^0 such that $u^0 \leq A(u^0)$. Use method 1 with jumps until z^n becomes small relative to $\| A(u^n) - u^n \|$, and then switch to method 2 or 3.

SUMMARY

This chapter completes our introduction to dynamic programming. States, transitions, and functional equations played key roles here, as in previous chapters. Linear programming, policy iteration, successive approximation, and turnpike theorems were introduced here. And a glimpse was provided of the methods and techniques that are the subject of our companion volume.

BIBLIOGRAPHIC NOTES

The model analyzed here is a special case of the two-player stochastic game analyzed in a classic paper by Shapley (1953). A lovely book by Howard (1960) highlighted policy iteration and aroused such great interest in this model that we can only cite a few subsequent references. The intimate relationship between (linear) Program 2 and policy iteration was first observed by deGhellinck (1960). As concerns successive approximation, Shapley and Howard studied method 1, Hastings (1968) seems to have introduced method 2, and Reetz (1971, 1973) and Totten (1971) seem to have introduced method 3. Van Nunen and Wessels (1977) unified and extended these three methods. MacQueen (1966) employed jumps with method 1, and White (1963) introduced related ideas for an undiscounted model. For an early turnpike theorem, see Shapiro (1968). Fox (personal communication) alerted this writer to the information in Exercise 11. Condition (*) specializes to this model a monotonicity property developed in this writer's dissertation (1965) and published jointly with Mitten, his advisor [Denardo and Mitten (1967)]

Of the many other significant contributions to Markovian decision models, we cite work of De Leve (1964), Blackwell (1965), Derman (1965), Schweitzer (1968), Hinderer (1970), Hordijk (1974), Van Nunen (1976) Federgruen (1978), Van Hee (1978), Kallenberg (1980), this writer (1968a).

PROBLEMS

1. (*Newton–Raphson iteration*) Let $\mathbb{R}^N$ denote the set of $N \times 1$ vectors, and let E map $\mathbb{R}^N$ into itself. We seek a zero of E [i.e., an element v of $\mathbb{R}^N$ for which $0 = E(v)$]. (This "0" is the $N \times 1$ vector of zeros.) For each v in $\mathbb{R}^N$, let $\nabla E(v)$ be the $N \times N$ (Jacobian) matrix of partial derivatives; specifically,

$$[\nabla E(v)]_{ij} = \frac{\partial [E(v)]_i}{\partial v_j}$$

Newton–Raphson iteration is as follows. One starts an iteration with a fixed element x of $\mathbb{R}^N$. One approximates $E(v)$ by the linear function $F_x(v)$ whose value at x is $E(x)$ and whose matrix of slopes is $\nabla E(x)$:

$$F_x(v) = E(x) + \nabla E(x) \cdot (v - x)$$

One computes the zero y of $F_x(\cdot)$; that is, one moves from x^n to x^{n+1}, where

$$0 = E(x^n) + \nabla E(x^n) \cdot (x^{n+1} - x^n)$$

a. To apply Newton–Raphson iteration to our model, set

$$[E(v)]_i = -v_i + \max_k \left\{ R_i^k + \sum_{j=1}^N \alpha P_{ij}^k v_j \right\}$$

Show that $E(v) = 0$ if and only if $v = f$.

b. Ignoring "ties," show that for each v in R^N there exists a policy δ (that can vary with v) such that

$$\nabla E(v) = \alpha P^\delta - I$$

c. Show that the application of Newton–Raphson iteration to this model is policy iteration.

2. [MacQueen (1967)] Define A by (8-64) and (8-65). With v fixed, let $d = \min_i \{[A(v)]_i - v_i\}$, $D = \max_i \{[A(v)]_i - v_i\}$, $v(d) = v + \mathbf{1}d/(1 - \alpha)$, and $v(D) = v + \mathbf{1}D/(1 - \alpha)$.

a. Show that $v(d) \le f \le v(D)$.

[*Hint:* Show that $A[v(d)] \ge v(d)$; then use Theorem 10.]

b. With v still fixed, define y_i^k by

$$y_i^k = R_i^k + \alpha \sum_{j=1}^N P_{ij}^k v_j - v_i$$

Show that no optimal policy π can have $\pi(i) = k$, when y_i^k satisfies

$$y_i^k < \frac{d - \alpha D}{1 - \alpha}$$

[*Hint:* Show that $R_i^k + \sum_j \alpha P_{ij}^k v(D)_j < v(d)_i$, then that $R_i^k + \sum_j \alpha P_{ij}^k f_j < f_i$.]

  c. Show how to use part b to exclude decisions from further consideration when executing successive approximation, policy iteration, and linear programming.

[*Remark:* This *decision exclusion* method has the (potentially small) term $1 - \alpha$ in the denominator.]

3. With v^n defined by (8-11) and $v^0 = 0$, let $e^n = v^n - f$.

  a. With s_i^k defined by (8-7), verify that

$$e_i^n = \max_k \left\{ -s_i^k + \sum_j \alpha P_{ij}^k e_j^{n-1} \right\}$$

Interpret e^n as the reward vector for an n-epoch Markov decision problem with nonpositive rewards and with a terminal reward vector, $-f$.

  b. (*Decision exclusion*) Define π^n by (8-50). Set $e_+^n = \max_i \{e_i^n\}$ and $e_-^n = \min_i \{e_i^n\}$. Suppose that $s_i^k > \alpha(e_+^n - e_-^n)$. Show that $\pi^m(i) \neq k$ for all $m > n$. (This lets us exclude decision k from computation of e^m for $m > n$.)

Remark: If f has been found by a method other than successive approximation, the above decision-exclusion rule expedites the calculation of optimal *finite-horizon* policies.

Note: Methods akin to those in Problems 2 and 3 can be found in Hastings and Mello (1973), Kushner and Kleinman (1971), and Porteus (1975).

4. This problem concerns the Markov decision model whose data are given in the following table. Let $\alpha = 0.9$.

State, i	Decision, k	Reward, R_i^k	Transition Probabilities P_{i1}^k	P_{i2}^k
1	a	4	$\frac{1}{3}$	$\frac{2}{3}$
1	b	2	$\frac{2}{3}$	$\frac{1}{3}$
2	a	0	1	0
2	b	2	0	1

  a. Find f, the numbers s_i^k defined in (8-7) and an optimal policy.

  b. For each n, find π^n that satisfies (8-50).

[**Hint:** You may find it easier to use the results of Problem 3, with $e^0 = -f$.]

5. (*Transient matrices*) A square matrix Q is called *transient* if each entry of Q^n approaches 0 as $n \to \infty$.

  a. Show that Theorem 1 holds for any transient matrix Q, not just for $Q = \alpha P$ with $\alpha < 1$.

  b. Drop the specific form $Q^\delta = \alpha P^\delta$; instead, assume, for each δ, that Q^δ is nonnegative and transient. Do Theorems 2 through 5 remain valid? Does Theorem 8?

6. [Fox (1973b)] On a digital computer, multiplication takes significantly longer then addition. If successive approximation is used to solve the policy evaluation equation, the N^2 multiplications entailed in computing Pv^n constitute most of the time per iteration. The ith element of Pv^n is the inner product of the ith row of P and v^n. Let $g = (g_1, g_2)$ and $h = (h_1, h_2)$, where g_i and h_i are $1 \times n$ for $i = 1, 2$. The inner product $g \cdot h = g_1 h_1 + g_2 h_2$. Winograd (1968) observed that this inner product can be computed in other ways, including

$$g \cdot h = (g_1 + h_2) \cdot (g_2 + h_1) - g_1 \cdot g_2 - h_2 \cdot h_1$$

a. *Other* than computing $g_1 \cdot g_2$ and $h_2 \cdot h_1$, how many multiplications and how many additions are required to compute $g \cdot h$ by the foregoing method?

b. Consider the successive approximation scheme given by (8-11). Winograd's method for computing inner products can be used to decrease the number of multiplications at the expense of increasing the number of additions. How? How many multiplications and additions are required per iteration?

[**Hint:** Compute some things once, others once per iteration.]

c. Adapt Winograd's method to successive approximation methods 2 and 3.

d. Use Winograd's method to accelerate pricing in the revised simplex method.

e. Let A and B be $N \times N$ matrices. The ijth element of the matrix AB is an inner product. Use Winograd's method to accelerate computation of AB.

DATA STRUCTURES*

* This supplement is used only in the starred sections of this volume.

INTRODUCTION

A way of organizing data is called a *data structure*. Tables and lists are data structures. Many computer programs spend considerable fractions of their running times extracting records from storage, altering them, and refiling them. Data structures that expedite this make for fast programs.

The data structures that are used in this volume are described here. These data structures have many other uses in operations research and elsewhere. They are well worth knowing.

HEAPS

Consider an iterative procedure that has the following two properties. At each iteration, the largest of a set $\{Y_1, \ldots, Y_N\}$ of numbers must be extracted. Of these numbers, only r are changed at each iteration, r being much smaller than N. It takes $N - 1$ comparisons to extract the largest at the first iteration. However, we shall see that subsequent iterations take work proportional to $r \log_2 N$, at worst, when the numbers are kept in a "heap."

Let $\{Y_1 \ldots, Y_N\}$ be a set of N numbers. Assume that N is odd. (This can be accomplished, if need be, by adding one extra number to the set.) The numbers $Y_1, \ldots, Y_N$ are said to form a *heap* if

(S1-1) $$Y_i \geq \max \{Y_{2i}, Y_{2i+1}\} \qquad \text{for } 1 \leq i < N/2$$

A heap of five numbers is shown in Figure S1-1, with Y_i adjacent to "node" i; in this example, Y_1 is 16 and is adjacent the node (circle) with 1 inside. Nodes $2i$

FIGURE S1-1. A heap

FIGURE S1-1. A heap

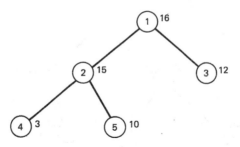

and $2i + 1$ are called the *sons* of node i. Node i is the *father* of nodes $2i$ and $2i + 1$. The sons of node 2 are nodes 4 and 5. Inequality (S1-1) assures that the label Y_i is as large or larger than the labels at both of i's sons. Note that Y_1 is the largest label. The next largest of the labels could either be Y_2 or Y_3.

How can we sort numbers into a heap? By induction. Assume that

$$(S1-2) \qquad Y_i \geq \max \{Y_{2i}, Y_{2i+1}\} \qquad \text{for } n \leq i < N/2$$

When $n = 1$, condition (S1-2) becomes condition (S1-1). When $n > N/2$, condition (S1-2) holds vacuously. The largest integer n for which (S1-2) might be violated is $(N - 1)/2$. The network in Figure S1-2 satisfies (S1-2) for $n = 3$, but not for $n = 2$. The difficulty is that 35 is smaller than the label at one of node 2's sons, namely label 50 at node 4. Interchange 35 with 50. Now 35 (at node 4) is smaller than the label at one of node 4's sons, namely label 40 at node 9. Interchange 35 with 40. The resulting network, shown in Figure S1-3, satisfies (S1-2) for $n = 2$. Label 35 has "filtered down" from node 2 to node 9. Inequality (S1-2) is not yet satisfied for the case $n = 1$. However, two more interchanges will filter label 37 down from node 1 to node 2 and then to node 4, at which point a heap will be formed. The scheme we have been describing is summarized below as the

FIGURE S1-2. Inequality (S1-2) holds for $n = 3$

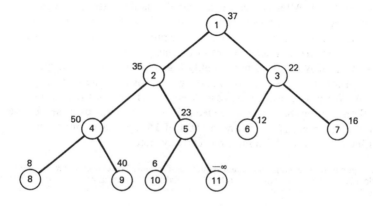

FIGURE S1-3. Inequality (S1-2) now holds for $n = 2$

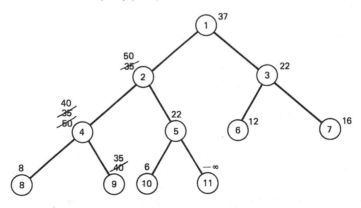

Heaping routine[1] (with N odd):

1. Set $n = (N + 1)/2$.
2. Let $n \leftarrow (n - 1)$. If $n = 0$, stop. Otherwise, set $j \leftarrow n$.
3. Set $s \leftarrow 2j$. If $Y_s < Y_{s+1}$, set $s \leftarrow (s + 1)$. If $Y_s > Y_j$, go to step 4; otherwise, go to step 2.
4. Interchange Y_s with Y_j. If $2s > N$, go to step 2. Otherwise, set $j \leftarrow s$, and go to step 3.

Exercise 1. Apply the heaping routine to the network in Figure S1-3.

> *Remark:* In some applications, one wishes to form a "heap" with the smallest number on top [i.e., with $Y_i \leq \min\{Y_{2i}, Y_{2i+1}\}$ in equation (S1-1)]. To do this, reverse the inequalities in step 3 of the heaping routine.

Suppose that a heap has been destroyed by altering one of its labels. If that label was increased, we can restore the heap by filtering the altered label upward, interchanging it with its father's label as often as necessary, until the heap is reformed. Alternatively, if the label was decreased, we can restore the heap by filtering it downward, interchanging it with a son's label, repeatedly, until the heap is restored. In either case, the number of interchanges is, at worst, one fewer than the number of rows in the heap.

A heap of N numbers has roughly $\log_2 N$ rows. One might guess that the number of interchanges needed to form a heap of N elements is approximately $N \log_2 N$. We now consider the case $N = 2^r - 1$ of r full rows and show that fewer than N interchanges are needed! (For the general case, see Problem 6.) When $r = 4$, we get $N = 15$, and a total of 15 nodes are found on four rows, which have 1, 2, 4, and 8 nodes, respectively. Let

---

[1] In an algorithm, a variable can take a sequence of values. The expression $x \leftarrow y$ means that variable x is to assume the value now taken by variable y. For instance, $n \leftarrow (n - 1)$ decreases n by 1.

$$f(r) = \text{the largest number of interchanges necessary to}$$
$$\text{put } 2^r - 1 \text{ numbers into a heap}$$

Of course, $f(1)$ equals 0, as no interchanges are needed to create a heap containing one number. Suppose that r exceeds 1. The key to computing $f(r)$ is the observation that node 2 and all nodes (such as 4, 5, 8, 9, etc.) pendant from node 2 form a system of $r - 1$ full rows. Similarly, node 3 and all nodes (such as 6, 7, 12, 13, etc.) pendant from it form a system of $r - 1$ full rows. Before the heaping routine deals at all with the number at node 1, it forms heaps out of these two subsets of nodes. It takes, *by definition*, at most $f(r - 1)$ interchanges to form each such subheap. To deal with the number at node 1 takes at most $r - 1$ additional interchanges. In short,

(S1-3)
$$f(r) = \begin{cases} 2f(r-1) + (r-1), & r \geq 2 \\ 0, & r = 1 \end{cases}$$

You may check, by substitution, that the solution to (S1-3) is

(S1-4)
$$f(r) = (2^r - 1) - r$$

Recall that $2^r - 1$ elements are being formed into a heap; (S1-4) indicates that this takes fewer than $2^r - 1$ interchanges.

HEAPS IN SORTING

One often wishes to sort records into sequence. Some of the fastest ways to do this are based on heaps. Suppose that one wishes to output the numbers $Y_1, \ldots, Y_N$ in sequence, largest first, then next largest, and so on. First form them into a heap; this takes at most N interchanges. Then iterate the following routine. Output the number at node 1, place $-\infty$ at node 1, and restart the heaping routine at step 2 with $n = 2$. Each iteration restores the heap by filtering the $-\infty$ downward. After N iterations, the heap will have been emptied out on sequence. Each iteration takes at most $\log_2 N$ interchanges. The total number of interchanges needed to output the heap in sequence is approximately $N \log_2 N$; an exact expression is given in Problem 7.

The sorting routine just described might be appropriate for a situation in which records are to be sorted and then filed in sequence. If the records are to be sorted and then processed in sequence, they would have to be read back into memory. Problem 5 sketches a slightly different method that sorts the records into sequence, keeps the records in memory, and requires only a few cells of storage in addition to those holding the records.

BUCKETS

Heaps are a fast and a general tool for maintaining a set of variables in such a way that the largest (or smallest) is readily found, as it is on the top of the heap. However, special-purpose methods can be faster when the data have structure.

Suppose that the numbers $Y_1, \ldots, Y_N$ are integers and that the difference between the smallest and largest of them is a small fraction of N. There will be a great many ties, and it may pay to assign a *bucket* to each possible value k that Y_n might take. In this application, bucket k would consist of a list of the variables whose value is currently k. As their values change, variables are shifted from one bucket to another, with, perhaps, less effort than is necessary to maintain a heap. Even if the numbers $Y_1, \ldots, Y_N$ are not integers, it may still pay to round down and assign a bucket k to each possible value[2] of $\lfloor Y_n \rfloor$, perhaps maintaining the lowest-numbered bucket as a heap and the others as lists.

LISTS

Within a computer, a bucket takes the form of a list, and records are moved from bucket to bucket (list to list) as computation progresses. This raises the issue of how to maintain a collection of lists within the computer's memory bank. The simplest sort of list consists of a block of contiguous memory locations. Unfortunately, the a priori bounds on the lengths of the lists are usually so long that insufficient computer memory is available to store each list in its own block. Occasionally, it is efficient to establish blocks of contiguous memory locations as buckets, perhaps with overflow buckets. Even when there is enough memory, blocks make it difficult to remove items from the interior of the lists.

An alternative is to store each item (record) in a fixed memory location and to keep track of the lists by using pointers. A *pointer* is the address at which an item (or record) is stored. Well-designed pointers let us insert and remove items from lists without scanning the lists. The fewer pointers the better, because they take space and must be kept current. We shall see that good pointer design depends on the sequence in which items are to be inserted and removed from the lists.

In a FIFO (first in, first out) list, the item inserted first on the list is to be removed first. Handle a FIFO list as follows. Keep a *head-of-list* pointer that records the memory address of the item inserted most recently on the list. Keep the list itself by using one pointer per item, in "circular" fashion. Specifically, except for the item inserted most recently, accompany each item on the list by a pointer to the memory address of the item inserted just after it. Accompany the item inserted most recently by a pointer to the memory address of the item inserted first.

This data structure is depicted in Figure S1-4, with items shown as nodes and pointers as directed arcs. Figure S1-4a depicts a list of five items, A through E, that were inserted in alphabetical sequence. The head-of-list pointer points to the memory location of item E, the one inserted most recently. Each item except E has a pointer to the one inserted just after it [e.g., item (node) C has a pointer to item (node) D].

[2] Throughout, $\lfloor x \rfloor$ is the largest integer that does not exceed x (e.g., $\lfloor 1 \rfloor = \lfloor 1.5 \rfloor = 1$).

FIGURE S1-4. Data structure for maintaining a FIFO list: (a) items *A* through *E* inserted in alphabetical sequence; (b) the list after item *A* (the first in) is removed; (c) the six-item list after item *F* (the new last in) is inserted

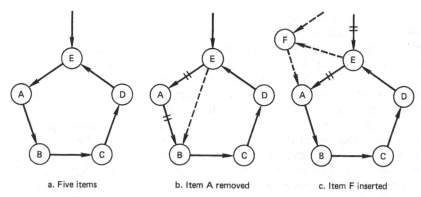

a. Five items b. Item A removed c. Item F inserted

Let us consider how to find and remove item *A* (the first in) from the list depicted in Figure S1-4a. Use the head-of-list pointer to find the memory address of item *E* (the last in), use item *E*'s pointer to find item *A*'s memory address, and reset item *E*'s pointer to the memory address in item *A*'s pointer, all of which is indicated by the dashed lines in Figure S1-4b.

Now consider how to insert item *F* (as the new last in) on the list depicted in Figure S1-4a. Use the head-of-list pointer to find the memory address of item *E* (the current last in), set item *F*'s pointer to the memory address in item *E*'s pointer, and then set item *E*'s pointer and the head-of-list pointer to the memory address of item *F*, all of which is indicated by the dashed lines in Figure S1-4c.

These pointers allow us to keep a FIFO list as a *singly linked* list, each item being accompanied by a single pointer. With one exception, each item's pointer points to its *successor*, which is the item inserted after it. This FIFO list is maintained without having to scan it (e.g., the insertions and deletions depicted in Figure S1-4 are accomplished without looking at items *B*, *C*, or *D*).

Exercise 2. Suppose that a FIFO list is maintained as indicated in Figure S1-4, except that the head-of-list pointer records the memory address of the item inserted first. What goes wrong?

Consider a LIFO (last in, first out) list. You should now suspect that a LIFO list can be kept as a singly linked list, with pointers to predecessors.

Exercise 3. Change some arrows in Figure S1-4 so as to allow maintenance of a LIFO list without scanning it.

LIFO and FIFO lists can be kept singly linked. To enable insertions and deletions in arbitrary positions without scanning, keep a *doubly linked* list, accompanying each item by a pointer to its predecessor and one to its successor.

HEAPS WITH POINTERS

In nearly every application of heaps, the number Y_i "at" node i of the heap is a portion of a record, not a complete record. Generally, Y_i is called the record's *key*.

Consider, for example, an inventory control system. Each item has its own record, which contains its natural-language description, its inventory level, its ordering information, and its unique identifying number (part number). Suppose that N such records are put in the computer's memory, and suppose it is desired to form them into a heap, by part-number sequence. This can be accomplished using several different data structures. One simple way is to leave the records where they are and to establish a separate block of N contiguous memory cells, these cells being labeled 1 through N. Cell i contains a pointer to the memory location of the record "at" node i of the heap. The record's part number (key) is kept with the record, *not* at cell i. To execute the heaping routine, compare keys and interchange pointers. At termination, node (cell) 1 contains the pointer to the memory location of the record with the largest key (part number).

BIBLIOGRAPHIC NOTES

Data structures are a subject of a brilliant and famous multivolume treatise by Knuth, the extant volumes being (1968, 1969, 1973). Another lucid source of information on data structures is Aho, Hopcroft, and Ullman (1974). Fox (1978) surveys the uses of data structures in operations research.

PROBLEMS

1. Suppose, in Figure S1-1, that the numbers Y_1 through Y_{11} at nodes 1 through 11 are, respectively, (3, 6, 8, 9, 10, 11, 12, 13, 14, 15, 16). Use the heaping routine to form them into a heap with the largest on top. Show all interchanges. How many interchanges were needed?

2. Alter the heaping routine to accommodate the case in which N can be even.

3. (*Alterations to heaps*) Suppose that the numbers $Y_1, \ldots, Y_N$ have been placed in a heap. Specify efficient routines that restore the heap after:

a. One number, Y_i, is increased.

b. One number, Y_j, is decreased.

c. One number, Y_k, is deleted.

[Hint: Switch Y_k with Y_N first.]

d. One number, Y_{N+1}, is added.

[Hint: Use part a.]

    e. Argue that $\log_2 N$ is a bound on the number of interchanges required by the routines in parts a, b, and c.

4. (*Sorting a heap*) Suppose that the numbers $Y_1, \ldots, Y_N$ form a heap, with N odd. The following is the part of a routine that outputs them in sequence. Output Y_1, then output the larger of Y_2 and Y_3, then move Y_N to node 1 and Y_{N-1} to node?, then $N \leftarrow (N - 2)$, and so on. Specify this routine completely.

5. (*Sorting a heap, continued*) Suppose that the numbers $Y_1, \ldots, Y_N$ form a heap, with N odd. Specify a routine that sorts them into sequence with the largest in position N, the next largest in position $N - 1$, and so on, but without using any external storage.

[Hint: Adapt Problem 4.]

6. Let $g(N)$ denote the number of interchanges needed to form a heap out of N numbers. We have seen that $g(2^r - 1) = 2^r - 1 - r$. Consider any remaining value of N (i.e., for some r, $2^{r-1} - 1 < N < 2^r - 1$).

    a. Suppose that $(2^{r-1} - 1) < N \leq (2^{r-1} - 1 + 2^{r-2})$. Hence, node 3 and those nodes pendant from node 3 form $(r - 2)$ complete rows. Show that

$$g(N) = (r - 1) + g(N - 2^{r-2}) + g(2^{r-2} - 1)$$
$$= g(N - 2^{r-2}) + 2^{r-2}$$

    b. Suppose that $(2^{r-1} - 1 + 2^{r-2}) < N < 2^r - 1$. Hence, node 2 and those nodes pendant from it form $(r - 1)$ complete rows. Show that

$$g(N) = g(N - 2^{r-1}) + 2^{r-1} - 1$$

    c. Show that $g(N) \leq N - 1$ for all N.

    Remark: Problem 6 verifies that fewer than N interchanges are needed to form a heap of N elements. When $N = 2^r$, the preceding bound is tight [i.e., $g(2^r) = 2^r - 1$].

7. Let $h(r)$ be the number of interchanges needed to output a heap of r full rows ($2^r - 1$ elements) by repeatedly replacing the element at node 1 by $-\infty$ and filtering the $-\infty$ down. At the end, each node has element $-\infty$ adjacent to it. The $-\infty$ adjacent a node in row n got there by $n - 1$ interchanges. Argue that

$$h(r) = \sum_{n=1}^{r} (n - 1)2^{n-1} = h(r - 1) + (r - 1)2^{r-1}$$
$$= 2[h(r - 1) + 2^{r-1} - 1]$$

[Hint: For the last expression, you might account for the subheaps pendant from nodes 2 and 3.]

Eliminate $h(r - 1)$ from the preceding pair of equations to get, with $N = 2^r - 1$,

$$h(r) = (N + 1)\log_2 (N + 1) - 2N$$

    Remark: Problem 7 verifies that the work needed to sort a heap of N elements into sequence is proportional to $N \log_2 N$ [i.e., $h(r)/(N \log_2 N)$ approaches 1 as $N \longrightarrow \infty$].

8. In an asychronous simulation, the "events" can occur at arbitrary times. Each *event* has an *occurrence time*, and the events that are known to occur at future times are kept on an *event list*. The simulation *clock* is advanced to the earliest of the occurrence times of the events on the event list, an event occurring at that time is removed from the event list, and that event is made to occur. Its occurrence may create one or more new events that are scheduled to occur at later times. Show how to keep the event list (a) as a FIFO list; (b) as a heap; (c) sorted into sequence by occurrence time. Have you a preference? Why?

9. (*Relates to Problem 8*) Consider a *synchronous* simulation, whose clock is advanced by fixed time increments. Describe a way to keep the event list in buckets. When might that be preferable?

10. The description in Figure S1-4 of a list management scheme is incomplete; it does not handle empty lists. Alter it to do so.

11. Adapt the heaping routine as suggested in the section on heaps with pointers, so that cell i contains a pointer to the memory location of the record that is currently "at" node i.

12. Reaching with buckets of width m (see the section on shortest paths in cyclic networks of Chapter 2) can be implemented with several data structures. Desirable implementations avoid scanning lists.

   a. Describe a data structure that associates with each node i the label v_i and two pointers.

[Hint: Keep each bucket as a doubly linked list.]

   b. Describe a data structure that keeps each bucket as a singly linked list, with a pointer from node i to the record (entry in a bucket) containing v_i.

   c. Account for the number of memory accesses and pointer adjustments needed for the data structures in parts a and b.

CONVEX FUNCTIONS*

* This supplement is used only in the starred sections of this volume.

INTRODUCTION

Convex functions play an important role in this volume, in operations research, and throughout mathematical analysis. The most pervasive aspects of convexity concern functions whose domain is the set of real numbers or a subset thereof. We study these functions first and turn briefly to other domains at this supplement's end.

Let $\mathbb{R}$ denote the set of real numbers. A subset of S of $\mathbb{R}$ is called an *interval* if $\alpha x + (1 - \alpha)y$ is in S for all x in S, all y in S, and all α having $0 < \alpha < 1$. An interval that contains two numbers must contain all numbers that lie between them. Examples of intervals are $\mathbb{R}$, the set of nonnegative numbers, and $\{x \mid 0 < x < 1\}$. An element x of an interval S is called an *interior point* of S if there exists a positive number ϵ such that every number y that satisfies $|y - x| < \epsilon$ is an element of S. Exactly one of the elements of the interval $S = \{x \mid 0 < x \leq 1\}$ is not an interior point. An interval is called an *open* interval if each of its elements is an interior point.

CHORDS AND CONVEXITY

Let g be a function that maps an interval S into the real numbers. Such a function g is called *convex on S* if for all x in S, y in S and $0 < \alpha < 1$,

(S2-1) $$g[\alpha x + (1 - \alpha)y] \leq \alpha g(x) + (1 - \alpha)g(y)$$

Figure S2-1 depicts a function g that is convex on $\mathbb{R}$. This figure highlights the values $g(x)$ and $g(y)$ that g assigns to x and y, as well as the *chord* (segment of a

FIGURE S2-1. A convex function g and one of g's chords

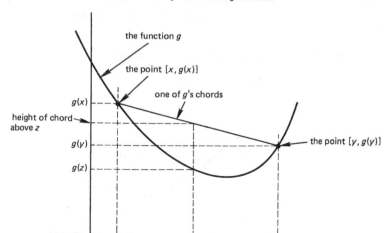

straight line) that connects the points $[x, g(x)]$ and $[y, g(y)]$ in the plane. Define z in terms of x, y, and α by

(S2-2)
$$z = \alpha x + (1 - \alpha)y = y + \alpha(x - y)$$

We note that z lies between x and y if and only if α lies between 0 and 1. For instance, when α is negative, the number $y + \alpha(x - y)$ is "reached" by "starting" at y and "moving" in the direction $-(x - y)$, which is away from x. This is true whether or not x lies below y. Similarly, the expression $\alpha g(x) + (1 - \alpha)g(y)$ takes values between $g(x)$ and $g(y)$ whenever α lies between 0 and 1. In the context of Figure S2-1, a function g is convex if it lies on or below each of its chords, that is, if the value $g(z)$ that the function assigns to $z = \alpha x + (1 - \alpha)y$ is no larger than the height $\alpha g(x) + (1 - \alpha)g(y)$ of the chord above z.

We shall see that a convex function need not be differentiable. Indeed, many of the functions encountered in this volume are convex, but not differentiable. So we first study the aspects of convexity that do not presume differentiability.

MONOTONE SLOPES AND CONVEXITY

Again, let g be a function that maps an interval S into $\mathbb{R}$. With dependence on the (fixed) function g suppressed, define

(S2-3)
$$s(x, y) = \frac{g(y) - g(x)}{y - x}, \quad \text{all } x, y \in S \text{ with } x \neq y$$

Interpret $s(x, y)$ as the *slope* of the chord connecting points $[x, g(x)]$ and $[y, g(y)]$ in the plane. A convex function g and its values at the points a, b, and c are depicted in Figure S2-2. The slopes of its chords are nondecreasing in the sense that $s(a, b) \leq s(b, c)$ whenever $a < b < c$. Figure S2-2 motivates the following lemma.

FIGURE S2-2. A convex function and two of its chords

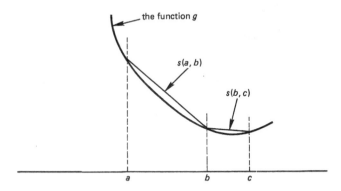

LEMMA 1. Let g map an interval S into $\mathbb{R}$. The function g is convex on S if and only if

(S2-4) $s(a, b) \leq s(b, c)$,   all $a, b, c \in S$ with $a < b < c$

Proof. Let a, b, and c satisfy $a < b < c$, and use the identity

(S2-5) $$b = \left(\frac{c - b}{c - a}\right)a + \left(\frac{b - a}{c - a}\right)c$$

to write $b = \alpha a + (1 - \alpha)c$, where $\alpha = (c - b)/(c - a)$ and where $0 < \alpha < 1$. Now suppose that g is convex on S. Apply (S2-1).

(S2-6) $$g(b) \leq \left(\frac{c - b}{c - a}\right)g(a) + \left(\frac{b - a}{c - a}\right)g(c)$$

Multiply (S2-6) by $(c - a)$, write $(c - a) = (c - b) + (b - a)$, and then rearrange terms as follows.

(S2-7) $$(c - b)[g(b) - g(a)] \leq (b - a)[g(c) - g(b)]$$

To obtain (S2-4), divide (S2-7) by $(b - a)(c - b)$.

Now suppose that g satisfies (S2-4). Reverse the argument; for example, first multiply (S2-4) by $(b - a)(c - b)$ to obtain (S2-7). Then manipulate (S2-7) into (S2-6), which verifies (S2-1) with $\alpha = (c - b)/(c - a)$.   ■

The next aspect of convex functions entails one new definition. Let T be any subset of $\mathbb{R}$, and let g map T into $\mathbb{R}$. Call g *nondecreasing* on T if $g(x) \leq g(y)$ for all x and y in T having $x < y$. Three simple properties of convex functions are given in the following lemma.

LEMMA 2. Let S be an interval and let x be an interior point of S. If g is convex on S, then:

    a. g is continuous at x.

    b. The limits in (S2-8) and (S2-9), below, exist.

(S2-8)
$$g'_+(x) = \lim_{\substack{\epsilon \to 0 \\ \epsilon > 0}} \{s(x, x + \epsilon)\}$$

(S2-9)
$$g'_-(x) = \lim_{\substack{\delta \to 0 \\ \delta < 0}} \{s(x + \delta, x)\}$$

    c. $-\infty < g'_-(x) \leq g'_+(x) < \infty$.

    d. The functions $g'_-(\cdot)$ and $g'_+(\cdot)$ are nondecreasing on the interior of S.

Remark: Recognize $g'_+(x)$ and $g'_-(x)$, respectively, as the *right* and *left* derivatives of g, evaluated at x. The lemma asserts that these (directional) derivatives exist, that they are nondecreasing on the interior of S, and that·the right derivative is at least as large as the left derivative. The lemma does *not* assert that a convex function is differentiable, and Problem 8 specifies a convex function having $g'_-(x) < g'_+(x)$ for every rational number x in its domain.

Proof. Since x lies in the interior of S, there exist $\delta < 0$ and $\epsilon > 0$ such that S contains $x + \delta$ and $x + \epsilon$. Apply Lemma 1 with $a = x + \delta, b = x$, and $c = x + \epsilon$ to get $s(x + \delta, x) \leq s(x, x + \epsilon)$. This inequality is preserved under the following limits.

$$-\infty < \limsup_{\substack{\delta \to 0 \\ \delta < 0}} s(x + \delta, x) \leq \liminf_{\substack{\epsilon \to 0 \\ \epsilon > 0}} s(x, x + \epsilon) < \infty$$

To see that $s(a, \cdot)$ and $s(\cdot, c)$ are nondecreasing, take $a < b < c$, and start with the identity $(c - a)s(a, c) = (b - a)s(a, b) + (c - b)s(b, c)$. Lemma 1 gives $s(a, b) \leq s(b, c)$, which combines with the preceding identity to give both halves of the inequality $s(a, b) \leq s(a, c) \leq s(b, c)$. This verifies that $s(a, \cdot)$ and $s(\cdot, c)$ are nondecreasing. Hence, "lim inf" and "lim sup" in the displayed equation can be replaced by "lim." This verifies parts (b) and (c).

A function that has right and left derivatives at x must be continuous at x; this verifies part (a).

Two applications of Lemma 1 give $s(x + \delta, x) \leq s(x, y) \leq s(y, y + \epsilon)$ where $x < y, \delta < 0$, and $\epsilon > 0$. Take limits to get $g'_-(x) \leq s(x, y) \leq g'_+(y)$. Since $s(x, \cdot)$ is nondecreasing, $g'_+(x) \leq g'_+(y)$. Similarly, since $s(\cdot, y)$ is nondecreasing, $g'_-(x) \leq g'_-(y)$. This proves part (d). ∎

Lemma 2 does not assert that a convex function is continuous at the end points of its domain. That it can jump upward is indicated by

Exercise 1. Let $g(x) = 0$ for $|x| < 1$ and $g(x) = 1$ for $|x| \geq 1$. Show that g is convex on the interval $S = \{x \,|\, -1 \leq x \leq +1\}$.

SUPPORTS AND CONVEXITY

The next characterization of convex functions entails the notion of a support. Let f and g be functions that map on interval S into $\mathbb{R}$. Call f *linear* on S if there exist real numbers a and b such that $f(x) = a + bx$ for all x in S. Call a linear function f a *support of g at x* if $f(x) = g(x)$ and if $f(y) \leq g(y)$ for every y in S. Figure S2-3 suggests (correctly) that a convex function has at least one support at each point in the interior of its domain.

FIGURE S2-3. The function g (in solid lines) and three of its supports (in dashed lines)

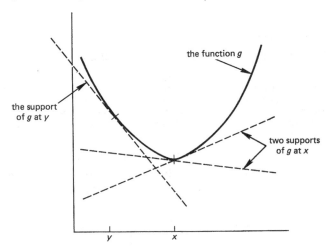

Exercise 1 (**continued**): Does the function g in Exercise 1 have supports at the points $x = -1$ and $x = 1$?

Exercise 1 suggests that convexity is more easily demonstrated for functions whose domains are open intervals.

LEMMA 3. Let g map an open interval S into $\mathbb{R}$. The following are equivalent.

    a. g is convex on S.

    b. There exists a real-valued function h that is nondecreasing on S and that satisfies

(S2-10) $$g(y) - g(x) = \int_x^y h(z)\, dz, \qquad \text{all } x, y \in S$$

    c. There exists a function f mapping S into $\mathbb{R}$ such that

(S2-11) $$g(y) \geq g(x) + (y - x)f(x), \qquad \text{all } x, y \in S$$

Remark: Part (b) is that g is the integral of a nondecreasing function h. Part (c) is that, for each x, the linear function L_x of y given by $L_x(y) = g(x) + (y - x)f(x)$ is a support of g at x [i.e., $g(x) = L_x(x)$ and $g(y) \geq L_x(y)$ for all $y \in S$]. Figure S2-3 suggests, correctly, that when g is differentiable, Lemma 3 holds for $h(x) = f(x) = g'(x)$.

Proof. It is shown that $(a) \Rightarrow (b) \Rightarrow (c) \Rightarrow (a)$.

$(a) \Rightarrow (b)$: Suppose that part (a) holds. Set $h(x) = g'_+(x)$. Part (d) of Lemma 2 shows that h is nondecreasing on the open interval S. Since h is nondecreasing, its ordinary (Riemann) integral exists (see Problem 6 for details). Since h is the right derivative of g, (S2-10) holds.

$(b) \Rightarrow (c)$: Suppose that part (b) holds. Consider $x < y$. Since h is nondecreasing,

$$g(y) - g(x) = \int_x^y h(z)\, dz \geq (y - x)h(x)$$

A similar argument holds when $x > y$; so (S2-11) holds.

$(c) \Rightarrow (a)$: Suppose that part (c) holds. Pick $a < b < c$, all in S. To get $s(b, c) \geq f(b)$, apply (S2-11) with $x = b$ and $y = c$. To get $s(a, b) \leq f(b)$, apply (S2-11) with $x = b$ and $y = a$. Hence, $s(a, b) \leq s(b, c)$, and Lemma 1 shows that g is convex. ∎

DIFFERENTIABILITY AND CONVEXITY

We now turn to the question of whether a differentiable function is convex. Let S be an open interval; the function g from S to $\mathbb{R}$ is called *differentiable on S* if its left and right derivatives exist and are identical on S [i.e., if $g'_-(x) = g'_+(x)$ for each x in S]. With T as a subset of $\mathbb{R}$, the function g from T to $\mathbb{R}$ is called *nonnegative on T* if $g(x) \geq 0$ for all x in T.

LEMMA 4. Let g map an open interval S into $\mathbb{R}$.

   a. If g is differentiable on S, the following are equivalent :

      (1) g is convex on S.

      (2) g' is nondecreasing on S.

   b. If g is twice differentiable on S, the following are equivalent:

      (1) g is convex on S.

      (2) g'' is nonnegative on S.

Remark: Lemma 4 asserts, among other things, that a twice differentiable convex function has a nondecreasing first derivative and a nonnegative second derivative.

Proof. First suppose that g is differentiable. Take $h = g'$, and note that part (a) is immediate from Lemma 3. Next, suppose that g is twice differentiable (i.e., g'' exists). Of course, $g'(y) - g'(x) = \int_x^y g''(z)\, dz$. Hence, g' is nondecreasing on S if and only if g'' is nonnegative on S. So part (b) is immediate from part (a). ∎

The simplest tests for convexity consist of computing derivatives and checking signs. Consider

Exercise 2. Find the largest domains on which each of the following functions are convex: x^2, x^3, e^x, e^{x^2}, $-\log x$.

COMPOSITES OF CONVEX FUNCTIONS

Suppose that g and h are convex functions on $\mathbb{R}$. It need not be the case that $f(x) = g[h(x)]$ is convex, as is witnessed below.

Exercise 3. Let $g(x) = -x$ and $h(x) = x^2$. Of the following, which are convex on $\mathbb{R}$: $g(x)$, $h(x)$, $g[h(x)]$?

In the following lemma, we see that the difficulty in Exercise 3 stems from the fact that g is not nondecreasing.

LEMMA 5. Let h be a convex function on the interval S, and let g be a convex function on an interval T that contains $\{h(x):x \in S\}$. If g is nondecreasing on T, then $g[h(x)]$ is convex on S.

   Proof. Since h is convex,

$$h[\alpha x + (1 - \alpha)y] \le \alpha h(x) + (1 - \alpha)h(y)$$

Apply to the above the fact that g is nondecreasing on T, then the fact that g is convex on T.

$$gh[\alpha x + (1 - \alpha)y] \le g[\alpha h(x) + (1 - \alpha)h(y)] \le \alpha gh(x) + (1 - \alpha)gh(y)$$

This demonstrates that the composite function gh is convex on S.   ∎

   Our attention now turns to sums of convex functions. Let g and h be convex on the interval S, and let c and d be nonnegative numbers. That the function $f(x) = cg(x) + dh(x)$ is convex on S is immediate from (S2-1). This extends to finite sums and, whenever the limits exist, to infinite sums. If, for instance, g is convex on $\mathbb{R}$ and if D is a real-valued random variable, the function f given by $f(x) = Eg(x + D)$ is convex on $\mathbb{R}$ whenever the expectation is finite.

Exercise 4. Let $(y - D)^+ = \max\{0, (y - D)\}$. Suppose that g is convex and nondecreasing on $\mathbb{R}_+ = \{x \in \mathbb{R} \,|\, x \ge 0\}$. Show that f is convex on $\mathbb{R}$, where

$$f(y) = 0.6g[(y - 1)^+] + 0.2g[(y - 2)^+] + 0.2g[(y - 3)^+]$$

INTEGERS AS A DOMAIN FOR CONVEXITY

Let J denote the set of nonnegative integers (i.e., $J = \{0, 1, \ldots\}$). Let $\mathbb{R}_+$ denote the set of nonnegative real numbers (i.e., $\mathbb{R}_+ = \{x \in \mathbb{R} \,|\, x \ge 0\}$). Let h be a real-valued function whose domain is J. We now probe the question: How can one determine whether h is the restriction to J of a function that is convex on

$\mathbb{R}_+$? Define the *first forward difference* Δh and the *second forward difference* $\Delta^2 h$ by

(S2-12) $\Delta h(i) = h(i + 1) - h(i)$,   all $i \in J$

(S2-13) $\Delta^2 h(i) = \Delta h(i + 1) - \Delta h(i)$,   all $i \in J$

Substitute (S2-12) into (S2-13) to get

(S2-14) $\Delta^2 h(i) = h(i + 2) - 2h(i + 1) + h(i)$,   all $i \in J$

The first and second forward differences are the discrete analogues of the first and second derivatives. Note from (S2-13) that Δh is nondecreasing on J if and only if $\Delta^2 h$ is nonnegative on J. One would guess (correctly) from Lemma 4 that h is the restriction to J of a convex function if Δh is nondecreasing on J or (equivalently) if $\Delta^2 h$ is nonnegative on J.

To extend h from J to $\mathbb{R}^+$ by *linear interpolation*, set

$$h(i + z) = h(i) + z\{h(i + 1) - h(i)\}$$

for all i in J and all z having $0 \le z < 1$.

LEMMA 6.   Let g be a convex function on $\mathbb{R}_+$. Then

(S2-15) $\Delta^2 g(i) \ge 0$,   all $i \in J$

Alternatively, let g be a function whose domain is J; if g satisfies (S2-15), then the extension of g to $\mathbb{R}_+$ by linear interpolation is convex on $\mathbb{R}_+$.

Proof.   Combine (S2-13), (S2-12), and (S2-3) to get

$$\Delta^2 g(i) = s(i + 1, i + 2) - s(i, i + 1),   \text{all } i \in J$$

Hence, Lemma 1 shows that a function g that is convex on $\mathbb{R}_+$ satisfies $\Delta^2 g(i) \ge 0$ for all $i \in J$.

Now assume that g has domain J and that g satisfies (S2-15). Define h on $\mathbb{R}_+$ by $h(x) = g(i + 1) - g(i)$ for $i \le x < i + 1$. Since $\Delta g(\cdot)$ is nondecreasing on J, $h(\cdot)$ is nondecreasing on $\mathbb{R}_+$. Extend g to $\mathbb{R}_+$ by linear interpolation, and note that $g(y) - g(x) = \int_x^y h(z)\,dz$. Hence, Lemma 3 shows that g is convex on $\mathbb{R}_+$.   ∎

With Lemma 6 in mind, we say that g is *convex on J* if (S2-15) holds. To verify that g is convex on J, one could compute the second forward difference. It sometimes suffices to calculate the second derivative of the natural extension of g to $\mathbb{R}_+$; see below.

Exercise 5.   Let $h(i) = (i + 1)^4 - (i + 1)^3$. Show   that $\Delta^2 h(i) \ge 0$ for all $i \in I$.
[Hint: Differentiate.]

More generally, if X is any subset of $\mathbb{R}$, we say that g is *convex on X* if g is the restriction to X of a function that is convex on an interval S that contains

X. Lemma 5 adapts to show that g is convex on the set of integers if and only $\Delta^2 g(x) \geq 0$ for every integer x.

CONCAVE FUNCTIONS

In each of the foregoing contexts, we call g a *concave* function if $-g$ is convex. So concave functions lie on or above their chords. Also, the first derivative of a concave function is nonincreasing and the second derivative is nonpositive, when they exist.

LARGER DOMAINS

At the close, we mention briefly convex sets and convex functions in higher-dimensional spaces. The space $\mathbb{R}^n$ is defined as the set of all n-tuples $x = (x_1, \ldots, x_n)$ of real numbers. If x and y are elements of $\mathbb{R}^n$ and α is any real number, then $\alpha x + y$ is the n-tuple whose ith element is $\alpha x_i + y_i$. A subset S of $\mathbb{R}^n$ is called a *convex set* if and only if the n-tuple $\alpha x + (1 - \alpha)y$ is in S whenever x and y are in S and whenever $0 \leq \alpha \leq 1$. Hence, the convex subsets of $\mathbb{R}^1$ are the intervals. A function g mapping a convex subset S of $\mathbb{R}^n$ into $\mathbb{R}$ is called *convex* on S if (S2-1) holds whenever x and y are in S and whenever $0 < \alpha < 1$.

Many of the properties of convex functions in higher-dimensional settings are suggested by the preceding lemmas. The analogues of Lemmas 2 to 4 can be shown to hold for a convex function g whose domain is an open convex subset S of $\mathbb{R}^n$. For instance, g is continuous on S, g has directional derivatives at each point x in S, and g has at least one support (supporting hyperplane) at each point x in S. If g is differentiable at x, then g has a unique support at x, and the directional derivatives of g match the slopes of this support. If g is not differentiable at x, then g has multiple supports at x.

We illustrate this for the case of a convex function whose domain is $\mathbb{R}^2$. A plane $p(\cdot)$ in $\mathbb{R}^2$ can be written $p(y_1, y_2) = a_0 + a_1 y_1 + a_2 y_2$, where a_0, a_1, and a_2 are constants. This plane is called a *support* of g at x if $p(x) = g(x)$ and $p(y) \leq g(y)$ for all y in $\mathbb{R}^2$. If g is not differentiable at x, it has multiple supports at x, as suggested by Figure S2-3. However, the partial derivatives of g in the positive x_1 and x_2 directions *need not* determine a support; see Problem 10 for an example of this.

Many of the applications of convexity that are found in this volume entail a function g that is convex on an interval in $\mathbb{R}$, but is *not* differentiable. One would like to generalize certain of these results to higher-dimensional settings. When attempting this, one often finds oneself blocked by the fact that directional derivatives fail to guarantee supports in higher dimensions. This block disappears when g is convex *and* differentiable, as in Theorem 6 of Chapter 5.

With the exception of Lemma 6, this supplement treats functions whose domains are intervals, not integers. One often models variables whose domains are integers (e.g., inventory levels are often integer-valued). Lemma 6 gives a simple test, namely (S2-15), that lets us transfer results back and forth between intervals and integers. It would be very desirable to have comparable tests for higher-dimensional settings. For instance, one would like a simple test as to whether or not a function g mapping $J \times J$ into $\mathbb{R}$ is the restriction to $J \times J$ of a function that is convex on $\mathbb{R}_+^2$. No simple test is known, and all such tests may be complex. The difficulty stems from the fact that linear interpolation of g is uniquely prescribed when its domain is J, but not when its domain is $J \times J$.

PROBLEMS

1. Let g be a convex function on an interval S of R. Show that g is convex on any interval T that is contained in S.

3. Set $S = \{x \,|\, x \geq 1\}$ and define g on S by $g(x) = e^x \log x$. Prove or disprove: g is convex on S.

3. Let g and h be convex functions on the interval S. Is $f(x) = \max \{g(x), h(x)\}$ convex on S? Need $p(x) = \min \{g(x), h(x)\}$ be convex on S?

4. Let g be a convex function on $S = \{x \,|\, x \geq 0\}$.
   a. Suppose that there exist x and y such that $0 < x < y$ and such that $g(x) < g(y)$. Show that $g(z) \longrightarrow \infty$ as $z \longrightarrow \infty$.
   b. Suppose that there exists exactly one $z^* > 0$ such that $g(z^*) = \min \{g(z) \,|\, z \geq 0\}$. Show that $g(z) \longrightarrow \infty$ as $z \longrightarrow \infty$.

5. Let g be a convex function on $S = \{x \,|\, 0 \leq x \leq 1\}$. Need parts b and c of Lemma 3 hold for g?
[Hint: Review Exercise 1.]

6. Let h be nondecreasing on $S = \{x \,|\, 0 \leq x \leq 1\}$. Let $0 = a_0 < a_1 < \ldots < a_N = 1$, with $a_i - a_{i-1} \leq \epsilon$ for each i. Show that $\sum_{i=0}^{N-1} (a_{i+1} - a_i)[h(a_{i+1}) - h(a_i)] \leq \epsilon[h(1) - h(0)]$. Show that h is Riemann-integrable on S.

7. Let h be convex on J. Suppose that $p_0, p_1, \ldots$ is a set of nonnegative numbers that sum to 1, and define f on J by

$$f(i) = \sum_{k=0}^{i} p_k h(i - k) + \sum_{k=i+1}^{\infty} p_k h(0)$$

Show that f is convex on J if $h(0) \leq h(1)$.
[Hint: Lemmas 5 and 6 help, with $(x)^+ = \max \{0, x\}$.]

8. (*Convexity and nondifferentiability*) Let $S = \{x \,|\, 0 < x < 1\}$. The set S contains countably many rational numbers; enumerate the rational numbers in S as $r_1, r_2, \ldots$. Define g on S by

$$g(x) = \sum_{i=1}^{\infty} 2^{-i}(x - r_i)^+$$

where $(z)^+ = \max \{0, z\}$.

a. Show that g is convex on S.

b. What points x have $g^-(x) < g^+(x)$?

9. Several famous inequalities from analysis are direct consequences of convexity. Let g be a convex function on an interval S, and let $\alpha_1, \ldots, \alpha_n$ be a set of nonnegative numbers whose sum is 1.

a. Prove *Jensen's inequality,*

$$g(\alpha_1 x_1 + \ldots + \alpha_n x_n) \le \alpha_1 g(x_1) + \ldots + \alpha_n g(x_n)$$

where $x_i \in S$ for $i = 1, \ldots, n$.

[**Hint:** Use induction on n.]

b. Let $S = \{x \mid x > 0\}$ and let $g(x) = -\log x$. Show that any set $x_1, \ldots, x_n$ of positive numbers satisfies

$$x_1^{\alpha_1} \ldots x_n^{\alpha_n} \le \alpha_1 x_1 + \ldots + \alpha_n x_n$$

(In other words, the *geometric mean* is no larger than the *arithmetic mean.*)

c. Use Jensen's inequality with $g(x) = x^p$ to demonstrate that when p exceeds 1 the set $x_1, \ldots, x_n$ of positive numbers satisfies

$$(\alpha_1 x_1 + \ldots + \alpha_n x_n)^p \le \alpha_1 x_1^p + \ldots + \alpha_n x_n^p$$

d. Let $w_1, \ldots, w_n$ be a set of positive numbers, let $\alpha = 1/p$, and let $\beta = 1 - \alpha$. Use part c with $\alpha_i = w_i/\sum w_j$ to obtain

$$\sum_i w_i x_i \le (\sum_i w_i)^\beta (\sum_i w_i x_i^{1/\alpha})^\alpha$$

e. *Hölder's inequality* is that

$$(\sum_i y_i z_i) \le (\sum_i y_i^{1/\beta})^\beta (\sum_i z_i^{1/\alpha})^\alpha$$

where $y_1, \ldots, y_n$ and $z_1, \ldots, z_n$ are sets of positive numbers, where $0 < \alpha < 1$, and where $\beta = 1 - \alpha$. Use part d with $y_i^{1/\beta} = w_i$ and $z_i^{1/\alpha} = ?$ to verify Hölder's inequality.

10. (*Directional derivatives and supports*) Let $g(x_1, x_2) = \max \{-x_1, -x_2\}$.

a. Show that g is convex on R^2.

b. Find a linear function h such that $h(0, 0) = g(0, 0)$ and such that $h(z) \le g(z)$ for all $z \in R^2$. [Such a function h is a support of g at $x = (0, 0)$.]

c. Compute the partial derivatives $g_1^+(0, 0)$ and $g_2^+(0, 0)$ of g with respect to x_1 and x_2, respectively, in the positive directions. Is the plane $h(z_1, z_2) = g(0, 0) + z_1 g_1^+(0, 0) + z_2 g_2^+(0, 0)$ a support of g at $x = (0, 0)$?

11. A function g is called *log convex* on an interval S if $\log g$ is convex on S.

a. Show that g is log convex on S if and only if

$$g[\alpha x + (1 - \alpha)y] \le [g(x)]^\alpha [g(y)]^{1-\alpha}$$

for all $x \in S$, $y \in S$, and $0 < \alpha < 1$.

b. Show that the function $g(x)h(x)$ is log-convex on the interval S if g and h are log convex on S.

c. More surprising is that the sum of two log-convex functions is log-convex. Suppose that $g(x)$ and $h(x)$ are log-convex on the interval S. Observe that

$$g[\alpha x + (1 - \alpha)y] + h[\alpha x + (1 - \alpha)y] \leq$$
$$[g(x)]^\alpha [g(y)]^{1-\alpha} + [h(x)]^\alpha [h(y)]^{1-\alpha}$$

Set $a = g(x)$, $b = g(y)$, $c = h(x)$, and $d = h(y)$. Use part b of Problem 11 twice, once with $x_1 = a/(a + c)$, to verify that

$$a^\alpha b^{1-\alpha} + c^\alpha d^{1-\alpha} \leq (a + c)^\alpha (b + d)^{1-\alpha}$$

and consequently, that $g(x) + h(x)$ is log-convex on S.

12. [Artin (1964)] Let $S = \{x \mid x > 0\}$. Define the *gamma function* Γ on S by

$$\Gamma(x) = \int_0^\infty e^{-t} t^{x-1} \, dt, \qquad x \in S$$

a. Show that the integral converges for $x > 0$.

b. Show that, for fixed t, the function $e^{-t} t^{x-1}$ is log-convex on S.

c. It is immediate from Problem 11 that finite sums of log-convex functions are log-convex. This extends to integrals when the sums converge. Show that Γ is log-convex on S.

13. (*Monotone failure rates*). Let T be a real-valued random variable having cumulative distribution function F and density function f [i.e., $F(t) = \Pr\{T < t\}$ and $f(t) = dF(t)/dt$ for all $t \in \mathbb{R}$]. The random variable T has *failure rate* function ϕ given by $\phi(t) = f(t)/[1 - F(t)]$. Let $S = \{t \mid 0 < F(t) < 1\}$. Show that the failure rate ϕ is nondecreasing (nonincreasing) on S if and only if the function $[1 - F(t)]$ is log-concave (log-convex) on S.

BIBLIOGRAPHY

AHO, A. V., J. F. HOPCROFT, and J. D. ULLMAN (1974), *The Design and Analysis of Computer Programs*, Addison-Wesley, Reading, Mass.

ARROW, K. J., D. BLACKWELL, and M. A. GIRSHICK (1949), "Bayes and Minimax Solutions of Sequential Decision Problems," *Econometrica*, 17, pp. 214–244.

ARROW, K. J., T. E. HARRIS, and J. MARSCHAK (1951), "Optimal Inventory Policy," *Econometrica*, 19, pp. 250–272.

ARTIN, E. (1964), *The Gamma Function*, Holt, Rinehart and Winston, New York. (Translation by Michael Batler of the 1931 German edition.)

BALINSKI, M. L., and H. P. YOUNG (1982), *Fair Representation*, Yale University Press, New Haven, Conn.

BEALE, E. M. L. (1959), "An Algorithm for Solving the Transportation Problem When the Shipping Cost over Each Route Is Convex," *Naval Research Logistics Quarterly*, 6, pp. 43–56.

BECKMANN, M. (1961), "An Inventory Model for Arbitrary Interval and Quantity Distributions of Demands," *Management Science*, 8, pp. 35–57.

BELLMAN, R. E. (1952), "On the Theory of Dynamic Programming," *Proceedings of the National Academy of Sciences*, 38, pp. 716–719.

BELLMAN, R. E. (1957a), *Dynamic Programming*, Princeton University Press, Princeton, N.J.

BELLMAN, R. E. (1957b), "A Markovian Decision Process," *Journal of Mathematics and Mechanics*, 6, pp. 679–684.

BELLMAN, R. E. (1958), "On a Routing Problem," *Quarterly of Applied Mathematics*, 16, pp. 87–90.

BELLMAN, R. E. (1971), *Introduction to the Mathematical Theory of Control Processes*, Volume 2, *Nonlinear Processes*, Academic Press, New York.

BELLMAN, R. E., AND S. E. DREYFUS (1962), *Applied Dynamic Programming*, Princeton University Press, Princeton, N.J.

BELLMAN, R., I. GLICKSBERG, AND O. GROSS (1955), "On the Optimal Inventory Equation," *Management Science*, 2, pp. 83–104.

BELLMORE, M., H. J. GREENBERG, AND J. J. JARVIS (1970), "Generalized Penalty-Function Concepts in Mathematical Optimization," *Operations Research*, 18, pp. 229–252.

BLACKWELL, D. (1965), "Discounted Dynamic Programming," *Annals of Mathematical Statistics*, 36, pp. 226–235.

BRADLEY, G. H., G. G. BROWN, AND G. W. GRAVES (1977), "Design and Implementation of Large Scale Transshipment Algorithms," *Management Science*, 24, pp. 1–34.

BROOKS, R., AND A. M. GEOFFRION (1966), "Finding Everett's Lagrange Multipliers by Linear Programming," *Operations Research*, 14, pp. 1149–1153.

CHARNES, A., AND W. W. COOPER (1961), "Multicopy Traffic Network Models, in *Theory of Traffic Flow*," R. HERMAN, ed., ELSEVIER, Amsterdam, pp. 84–96.

ÇINLAR, E. (1972), "Superposition of Point Processes," in *Stochastic Point Processes: Statistical Analysis, Theory, and Applications*, P. A. W. LEWIS, ed., Wiley-Interscience, New York.

ÇINLAR, E. (1975), *Introduction to Stochastic Processes*, Prentice-Hall, Englewood Cliffs, N. J.

DAFERMOS, S. C., AND F. T. SPARROW (1969), "The Traffic Assignment Problem for a General Network," *Journal of Research of the National Bureau of Standards*, 73B, pp. 91–118.

DANTZIG, G. B. (1957), "Discrete-Variable Extreme Problems," *Operations Research*, 5, pp. 268–273.

DEGHELLINCK, G. (1960), "Les Problèmes de décisions séquentielles," *Cahiers du Centre de Recherche Opérationnelle*, 2, pp. 161–179.

DEGHELLINCK, G., AND G. D. EPPEN (1967), "Linear Programming Solutions for Separable Markovian Decision Problems, "*Management Science*, 13, pp. 371–394.

DE LEVE, G. (1964), *Generalized Markovian Decision Processes, Part I: Model and Method; Part II: Probabilistic Background*, Mathematical Centre Tracts Nos. 3 and 4, Amsterdam.

DENARDO, E. V. (1965), "Sequential Decision Processes," Ph.D. dissertation, Northwestern University, Evanston, Ill.

DENARDO, E. V. (1968a), "Contraction Mappings in the Theory Underlying Dynamic Programming," *SIAM Review*, 9, 165–177.

DENARDO, E. V. (1968b), "Separable Markovian Decision Problems," *Management Science*, 14, pp. 451–462.

DENARDO, E. V., AND B. L. FOX (1979a), "Shortest-Route Methods: 1. Reaching, Pruning, and Buckets," *Operations Research*, 27, pp. 161–186.

DENARDO, E. V., AND B. L. FOX (1979b), "Shortest-Route Methods: 2. Group Knapsacks, Expanded Networks, and Branch-and-Bound," *Operations Research*, 27, pp. 548–566.

DENARDO, E. V., AND B. L. FOX (1980), "Enforcing Constraints on Expanded Networks," *Operations Research*, 28, pp. 1213–1218.

DENARDO, E. V., G. R. HUBERMAN, AND U. G. ROTHBLUM (1979), *Optimal Locations on a Line Are Interleaved*, Technical Report, Yale University, New Haven, Conn.

DENARDO, E. V., AND L. G. MITTEN (1967), "Elements of Sequential Decision Processes," *Journal of Industrial Engineering*, 18, pp. 106–111.

DERMAN, C. (1965), "Markovian Sequential Control Processes-Denumerable State Space," *Journal of Mathematical Analysis and Applications*, 10, pp. 295–302.

DIJKSTRA, E. W. (1959), "A Note on Two Problems in Connexion with Graphs," *Numerische Mathematik*, 1, pp. 269–271.

DREYFUS, S. E. (1966), *Dynamic Programming and the Calculus of Variations*, Academic Press, New York.

DREYFUS, S. E. (1969), "An Appraisal of Some Shortest-Path Algorithms," *Operations Research*, 17, pp. 395–412.

DREYFUS. S. E. AND A. LAW (1978), *The Art and Theory of Dynamic Programming*, Academic Press, New York.

DVORETSKY, A., J. KIEFER, AND J. WOLFOWITZ (1952a), "The Inventory Problem: I. Case of Known Distributions of Demand," *Econometrica*, 20, pp. 187–222.

DVORETSKY, A., J. KIEFER, AND J. WOLFOWITZ (1952b), "The Inventory Problem: II. Case of Unknown Distributions of Demand," *Econometrica*, 20, pp. 451–466.

ELMAGHRABY, S. E. (1973), "The Concept of 'State' in Discrete Dynamic Programming," *Journal of Mathematical Analysis and Applications*, 43, pp. 642–693.

ERICKSON, R. A. (1978), Minimum-Concave-Cost Single-Source Network Flows, Doctoral Dissertation, Department of Operations Research, Stanford University, Stanford, Calif.

EVERETT, H. (1963), "Generalized Lagrange Multiplier Method for Solving Problems of Optimum Allocation of Resources," *Operations Research*, 11, pp. 399–413.

FEDERGRUEN, A. (1978), *Markovian Control Problems: Functional Equations and Algorithms*, Mathematical Centre Tracts No. 97, Amsterdam.

FISHER, M. L. (1981), "Lagrangian Relaxation Method for Solving Integer Programming Problems," *Management Science*, 27, pp. 1–18.

FLORIAN, M., AND S. NGUYEN (1976), "An Application and Validation of Equilibrium Trip Assignment Methods," *Transportation Science*, 10, pp. 374–390.

FLOYD, R. W. (1962), "Algorithm 97, Shortest Path," *Communications of the Association for Computing Machinery*, 5, p. 345.

FONG, C. O., AND M. R. RAO (1973), "Capacity Expansion with Two Producing Regions and Concave Costs," *Management Science*, 22, pp. 331–339.

FORD, L. R., JR. (1956), *Network Flow Theory*, P-923, The Rand Corporation, Santa Monica, Calif.

FORD, L. R., JR., AND D. R. FULKERSON (1962), *Flows in Networks*, Princeton University Press, Princeton, N.J.

FOX, B. L. (1966), "Discrete Optimization via Marginal Analysis," *Management Science*, 13, pp. 210–215.

FOX, B. L. (1973a), "Calculating kth Shortest Paths," *INFOR*, 11, pp. 66–70.

Fox, B. L. (1973b), "Reducing the Number of Multiplications in Iterative Processes," *Acta Informatica*, 3, pp. 43–45.

Fox, B. L. (1978), "Data Structures and Computer Science Techniques in Operations Research," *Operations Research*, 26, pp. 686–717.

Fox, B. L., AND D. M. LANDI (1968), "An Algorithm for Identifying the Ergodic Subchains and Transient States of a Stochastic Matrix," *Communications of the Association for Computing Machinery*, 11, pp. 619–621.

Fox, B. L., AND D. M. LANDI (1970), "Searching for the Multiplier in One-Constraint Optimization Problems," *Operations Research*, 18, pp. 253–262.

GARFINKEL, R. S., AND G. L. NEMHAUSER (1972), *Integer Programming*, Wiley, New York.

GILMORE, P. C., AND R. E. GOMORY (1966), "The Theory and Computation of Knapsack Functions," *Operations Research*, 14, pp. 1045–1074.

GILSINN, J., AND C. WITZGALL (1973), *A Performance Comparison of Labeling Algorithms for Calculating Shortest Path Trees*, NBC Technical Note 772, National Bureau of Standards, Washington, D.C.

GLOVER, F., D. KARNEY, AND D. KLINGMAN (1974), "Implementation and Computational Comparisons of Primal, Dual and Primal–Dual Computer Codes for Minimum Cost Network Flow Problems," *Networks*, 4, pp. 191–212.

GROSS, O. (1956), *Notes on Linear Programming: Part XXX. A Class of Discrete-Type Minimization Problems*, RM-1644, The Rand Corporation, Santa Monica, Calif.

HASTINGS, N. A. J. (1968), "Some Notes on Dynamic Programming and Replacement," *Operational Research Quarterly*, 19, pp. 453–464.

HASTINGS, N. A. J., AND J. M. C. MELLO (1973), "Tests for Suboptimal Actions in Discounted Markov Programming," *Management Science*, 19, pp. 1019–1022.

HINDERER, K. (1970), "Foundations of Non-Stationary Dynamic Programming with Discrete Time Parameter," Lecture Notes in Operations Research and Mathematical Systems, edited by M. BECKMANN AND H. P. KUNZI, Springer-Verlag, Berlin.

HITCHNER, L. E. (1968), *A Comparative Investigation of the Computational Efficiency of Shortest Path Algorithms*, Report ORC 68–25, Operations Research Center, University of California, Berkeley, Calif.

HOFFMAN, A. J., AND S. WINOGRAD (1972), "Finding All Shortest Distances in a Directed Network," *IBM Journal of Research and Development*, 16, pp. 412–414.

HOFFMAN, W., AND R. PAVLEY (1959), "A Method for the Solution of the Nth Best Path Problem," *Journal of the Association for Computing Machinery*, 6, pp. 506–514.

HORDIJK, A. (1974), *Dynamic Programming and Markov Potential Theory*, Mathematical Centre Tracts No. 51, Amsterdam.

HOWARD, R. A. (1960), *Dynamic Programming and Markov Processes*, MIT Press, Cambridge, Mass.

HU, T. C. (1966), "Minimum-Cost Flows in Convex-Cost Networks," *Naval Research Logistics Quarterly*, 13, pp. 1–9.

HUNTINGTON, E. V. (1928), "The Apportionment of Representatives in Congress," *Transactions of the American Mathematical Society*, 30, pp. 85–110.

ISAACS, R. (1951), *Games of Pursuit*, P-257, The Rand Corporation, Santa Monica, Calif.

JOHNSON, D. B. (1973), "Algorithms for Shortest Paths," Ph.D. thesis, Department of Computer Science, Cornell University, Ithaca, N.Y.

KALLENBERG, L. C. M. (1980), *Linear Programming and Finite Markovian Control Problems*, Doctoral thesis, Mathematisch Centrum, Amsterdam.

KARLIN, S. (1960), "Dynamic Inventory Policy with Varying Stochastic Demands," *Management Science*, 6, pp. 231–258.

KHINCHINE, A. Y. (1960), *Mathematical Methods in the Theory of Queueing*, Charles Griffin, London.

KLINCEWICZ, J. C. (1979), "Algorithms for Network Flow Problems with Convex Separable Costs," Ph.D. dissertation, Yale University, New Haven, Conn.

KNUTH, D. E. (1968), *The Art of Computer Programming:* Vol. 1, *Fundamental Algorithms*, Addison-Wesley, Reading, Mass.

KNUTH, D. E. (1969), *The Art of Computer Programming:* Vol. 2, *Seminumerical Algorithms*, Addison-Wesley, Reading, Mass.

KNUTH, D. E. (1973), *The Art of Computer Programming:* Vol. 3, *Sorting and Searching*, Addison-Wesley, Reading, Mass.

KOLMOGOROV, A. N., AND FOMIN, S. V. *Elements of the Theory of Functions and Functional Analysis.* Vol. 1: *Metric and Normed Spaces*, Graylock Press, Rochester, N.Y. (1957 translation by L. F. BORON of the 1954 Russian edition),

KUSHNER, H. J., AND A. J. KLEINMANN (1971), "Accelerated Procedures for the Solution of Discrete Markov Control Problems," *IEEE Transactions on Automatic Control*, AC-16, pp. 147–152.

LARSON, R. E. (1968), *State Increment Dynamic Programming*, American Elsevier, New York.

LOVE, S. F. (1973), "Bounded Production and Inventory Models with Piecewise Concave Costs," *Management Science*, 20, pp. 313–318.

MACQUEEN, J. (1966), "A Modified Dynamic Programming Method for Markovian Decision Problems," *Journal of Mathematical Analysis and Applications*, 14, pp. 38–43.

MACQUEEN, J. (1967), "A Test for Suboptimal Actions in Markovian Decision Problems," *Operations Research*, 15, pp. 559–561.

MAGAZINE, M. J., G. L. NEMHAUSER, AND L. E. TROTTER (1975), "When the Greedy Solution Solves a Class of Knapsack Problems," *Operations Research*, 23, pp. 207–217.

MANNE, A. S. (1958), "Programming of Economic Lot Sizes," *Management Science*, 4, pp. 115–135.

MANNE, A. S., AND A. F. VEINOTT, JR. (1967), in *Investments for Capacity Expansion: Size, Location and Time-Phasing*, A. S. MANNE, ed., MIT Press, Cambridge, Mass., Chap. 11.

Massé, P. (1946), *Les Réserves et la régulation de l'avenir dans la vie économique*, 2 vols., Hermann, Paris.

Mitten, L. G., and T. Probhakar (1964), "Optimization of Batch-Reflux Processes by Dynamic Programming," *Chemical Engineering Progress, Symposium Series*, 60, pp. 53–59.

Moore, E. F. (1957), "The Shortest Path through a Maze," in *Proceedings of an International Symposium on the Theory of Switching*, Vol. 2 (*The Annals of the Computation Laboratory of Harvard University*, 30), Harvard University Press, Cambridge, Mass.

Morin, T. L., and R. E. Marsten (1976), "An Algorithm for Nonlinear Knapsack Problems," *Management Science*, 22, pp. 1147–1158.

Nguyen, S. (1974), "An Algorithm for the Traffic Assignment Problem," *Transportation Science*, 8, pp. 203–216.

Pollack, M., and W. Wiebson (1960), "Solutions of the Shortest-Route Problem—A Review," *Operations Research*, 8, pp. 224–230.

Porteus, E. L. (1971), "On the Optimality of Generalized (s, S) Policies," *Management Science*, 17, pp. 411–426.

Porteus, E. L. (1975), "Bounds and Transformations for Discounted Finite Markov Decision Chains," *Operations Research*, 23, pp. 761–764.

Potts, R. B., and R. M. Oliver (1972), *Flows in Transportation Networks*, Academic Press, New York.

Reetz, D. (1971), *Solution of a Markovian Decision Problem by Single Step Value Iteration*, Institut für Gesellschafts- und Wirtschaftswissenschaften, Universität Bonn, Bonn.

Reetz, D. (1973), "Solution of a Markovian Decision Problem by Overrelaxation," *Zeitschrift für Operations Research*, 17, pp. 29–32.

Rockafellar, R. T. (1972), *Convex Analysis*, 2nd printing, Princeton University Press, Princeton, N.J.

Scarf, H. (1960), "The Optimality of (S, s) Policies in the Dynamic Inventory Problem," in *Mathematical Methods in the Social Sciences* 1959, K. Arrow, S. Karlin, and P. Suppes, eds., Stanford University Press, Stanford, Calif., Chap. 13.

Scarf, H. (1963), "A Survey of Analytic Techniques in Inventory Theory," in *Multistage Inventory Models and Techniques*, H. Scarf, D. Gilford, and M. Shelly, eds., Stanford University Press, Stanford, Calif., Chap. 7.

Schäl, M. (1976), "On the Optimality of (s, S)-Policies in Dynamic Inventory Models with Finite Horizon," *SIAM Journal on Applied Mathematics*, 30, pp. 528–537.

Schweitzer, P. J. (1968), "Perturbation Theory and Finite Markov Chains," *Journal of Applied Probability*, 5, pp. 401–413.

Shapiro, J. F. (1968), "Turnpike Planning Horizons for a Markovian Decision Model," *Management Science*, 14, pp. 292–300.

Shapiro, J. F. (1979), "Nonlinearly Constrained Nonlinear Programming Problems: Generalized Lagrangean Functions and Penalty Methods," in *Mathematical Programming: Structures and Algorithms*, Wiley, New York, Sec. 7.4.

SHAPLEY, L. S. (1953), "Stochastic Games," *Proceedings of the National Academy of Sciences*, 39, pp. 1095–1100.

SHERBROOKE, C. C. (1968), "Metric: A Multi-echelon Technique for Recoverable Item Control," *Operations Research*, 16, pp. 122–141.

SPIRA, P. M. (1973), "A New Algorithm for Finding All Shortest Paths in a Graph of Positive Arcs in Average Time $O(n^2 \log^2 n)$," *SIAM Journal on Computing*, 2, pp. 28–32.

TIJMS, H. C. (1972), *Analysis of (s, S) Inventory Models*, Mathematical Centre Tracts No. 40, Amsterdam.

TOTTEN, J. C. (1971), *Computational Methods for Finite State Finite Valued Markovian Decision Problems*, Report 71–9, Operations Research Center, University of California, Berkeley, Calif.

VAN HEE, K. M. (1978), *Bayesian Control of Markov Chains*, Mathematical Centre Tracts No. 95, Amsterdam.

VAN NUNEN, J. A. E. E. (1976), *Contracting Markov Decision Processes*, Mathematical Centre Tracts No. 71, Amsterdam.

VAN NUNEN, J. A. E. E., AND J. WESSELS (1977), "The Generation of Successive Approximations for Markov Decision Processes by Using Stopping Times," in Mathematical Centre Tracts No. 93, H. C. TIJMS AND J. WESSELS, eds., Amsterdam, pp. 25–37.

VEINOTT, A. F., JR. (1964), "Production Planning with Convex Costs: A Parametric Study," *Management Science*, 10, pp. 441–460.

VEINOTT, A. F., JR. (1965), "Optimal Policy for a Multi-product, Dynamic Non-stationary Inventory Problem," *Management Science*, 12, pp. 206–222.

VEINOTT, A. F., JR. (1966a), "The Status of Mathematical Inventory Theory," *Management Science*, 12, pp. 745–777.

VEINOTT, A. F., JR. (1966b), "On the Optimality of (s, S) Inventory Policies: New Conditions and a New Proof," *SIAM Journal on Applied Mathematics*, 14, pp. 1067–1083.

VEINOTT, A. F., JR. (1973), "Inventory Theory," classnotes, Department of Operations Research, Stanford University, Stanford, Calif.

VEINOTT, A. F., JR., AND H. M. WAGNER (1965), "Computing Optimal (s, S) Inventory Policies," *Management Science*, 11, pp. 525–552.

WAGNER, H. M., AND T. WHITIN (1957), "Dynamic Problems in the Theory of the Firm," *Theory of Inventory Management*, 2nd ed., T. WHITIN, ed., Princeton University Press, Princeton, N.J., App. 6.

WALD, A. (1947), *Sequential Analysis*, Wiley, New York.

WARDROP, J. G. (1952), "Some Theoretical Aspects of Road Traffic Research," *Proceedings of the Inst. of Civil Engineers*, Part II, pp. 325–378.

WARSHALL, S. (1962), "A Theorem on Boolean Matrices," *Journal of the Association for Computing Machinery*, 9, pp. 11–12.

WHITE, D. J. (1963), "Dynamic Programming, Markov Chains and the Method of Successive Approximations," *Journal of Mathematical Analysis and Applications*, 6, pp. 373–376.

WINOGRAD, S (1968), "A New Algorithm for Inner Product," *IEEE Transactions Computers* C-17, pp. 693–694.

YEN, J. Y. (1970), "A Shortest Path Algorithm," Ph.D. dissertation, University of California, Berkeley, Calif.

ZADEH, N. (1979), *A Simple Alternative to the Out-of-Kilter Algorithm*, Technical Report No. 25, Department of Operations Research, Stanford University, Stanford, Calif.

ZANGWILL, W. I. (1966), "A Deterministic Multi-period Production Scheduling Model with Backlogging," *Management Science*, 13, pp. 105–119.

ZANGWILL, W. I. (1969), "A Backlogging Model and a Multi-echelon Model of a Dynamic Economic Lost Size Production System—A Network Approach," *Management Science*, 15, pp. 506–527.

INDEX

SUBJECT INDEX

NAME INDEX